REPORT

Air Force Contingency Contracting

Reachback and Other Opportunities for Improvement

John A. Ausink, Laura Werber Castaneda,
Mary E. Chenoweth

Prepared for the United States Air Force

RAND PROJECT AIR FORCE

The research described in this report was sponsored by the United States Air Force under Contract FA7014-06-C-0001. Further information may be obtained from the Strategic Planning Division, Directorate of Plans, Hq USAF.

Library of Congress Cataloging-in-Publication Data

Ausink, John A.
  Air Force contingency contracting : reachback and other opportunities for improvement / John A. Ausink, Laura Werber Castaneda, Mary E. Chenoweth.
      p. cm.
  Includes bibliographical references.
  ISBN 978-0-8330-5012-0 (pbk. : alk. paper)
  1. United States. Air Force—Procurement—Evaluation. 2. Defense contracts—United States—Evaluation. 3. Contracting out—United States—Evaluation. I. Castaneda, Laura Werber. II. Chenoweth, Mary E. III. Title.

  UG1123.A848 2011
  358.4'152120973—dc22

                                                                                    2011004166

The RAND Corporation is a nonprofit institution that helps improve policy and decisionmaking through research and analysis. RAND's publications do not necessarily reflect the opinions of its research clients and sponsors.

RAND® is a registered trademark.

Published 2011 by the RAND Corporation
1776 Main Street, P.O. Box 2138, Santa Monica, CA 90407-2138
1200 South Hayes Street, Arlington, VA 22202-5050
4570 Fifth Avenue, Suite 600, Pittsburgh, PA 15213-2665
RAND URL: http://www.rand.org/
To order RAND documents or to obtain additional information, contact
Distribution Services: Telephone: (310) 451-7002;
Fax: (310) 451-6915; Email: order@rand.org

Preface

Operations Iraqi Freedom and Enduring Freedom have placed great demands on the U.S. Air Force's (USAF's) highly skilled contracting workforce. In late 2008, acknowledging the reality of these demands, the Air Force established a deploy-to-dwell ratio of 1:1 for the contracting career field—military personnel are deployed for six months, come home for six months, and are then deployed again (Correll, 2008). The Office of the Deputy Assistant Secretary of the Air Force for Contracting (SAF/AQC) asked RAND Project AIR FORCE to examine "reachback"—the use of contracting capability outside of the theater of operations to accomplish contracting tasks for customers in theater—as a potential means for reducing the deployment burden on military personnel. In addition, reachback might improve performance in some areas because of greater personnel continuity, standardization of processes, and the ability to access personnel with higher-level skills.

This report documents the results of RAND's research on this topic and shows that, although reachback has the potential to reduce deployments and increase the effectiveness of some contracting functions, there are also other important issues the Air Force can address to decrease the stress on the contracting career field.

The research described in this report was conducted within the Resource Management Program of RAND Project AIR FORCE for the fiscal year (FY) 2009 study "Improving the Effectiveness of Contingency Contracting Through Reachback." It should interest military leadership or policymakers involved in the management or use of contingency contracting personnel, the analytical community that studies contingency contracting, proponents of the Air Force contracting career field, and contingency contracting officers themselves.

RAND Project AIR FORCE

RAND Project AIR FORCE (PAF), a division of the RAND Corporation, is the U.S. Air Force's federally funded research and development center for studies and analyses. PAF provides the Air Force with independent analyses of policy alternatives affecting the development, employment, combat readiness, and support of current and future aerospace forces. Research is conducted in four programs: Force Modernization and Employment; Manpower, Personnel, and Training; Resource Management; and Strategy and Doctrine.

Additional information about PAF is available on our website:
http://www.rand.org/paf/

Contents

Figures

Tables

Summary

The Air Force's contracting career fields for both officers and enlisted personnel are under stress because of increased demands placed on them by deployments for contingency operations in Iraq and Afghanistan. Evidence of stress includes

- a deploy-to-dwell ratio of 1:1, which currently means that contracting personnel can expect to deploy for six months, return home for six months, and then deploy again
- a heavy reliance on the Air Force for contingency contracting positions, as indicated by the fact that 70 percent of the positions in the Joint Contracting Command Iraq/Afghanistan (JCC-I/A)[1] are filled by Air Force contracting personnel
- the potential difficulty of retaining experienced, senior-level personnel in the career field, as indicated by a decrease in the average career length of enlisted personnel from over ten years in FY 2004 to under seven years in FY 2008, and a decline in the number of personnel in the "journeyman/craftsman" cohort of personnel with 11 to 15 years of experience.

One potential approach to decreasing deployment stress on the career field is to increase the use of "reachback" for some contracting functions. This means that tasks normally accomplished by deployed contingency contracting officers (CCOs) in theater are accomplished by personnel outside the theater of operations—perhaps in the continental United States (CONUS). For example, legal review of a contract might be done by a specialist in CONUS instead of the CCO in theater. This could mean that fewer people need to be deployed or the workload of those deployed could be decreased.

This report uses data from four sources—CCO after action reports (AARs), focus groups, interviews with subject-matter experts, and purchasing data recorded in the Joint Contingency Contracting System (JCCS)—to examine what CCOs do and to better understand some of the challenges they face while accomplishing their mission.[2]

Our initial intent was only to develop broad criteria for the use of reachback, examine JCC-I/A purchases that could be made using reachback, and estimate the potential deployment impact if reachback were used for certain categories of purchases. However, we learned that, while CCOs recognized the potential of reachback to reduce deployments, they felt it had more potential to improve other aspects of contingency contracting, such as continuity of workflow, standardization of contract requirements, concentration of contracting expertise in

---

[1] After this report was completed, CENTCOM established the CENTCOM Contracting Command, which is based in Qatar. JCC-I/A transitioned to the new organization in June 2010.

[2] We believe that our work with the AARs represents the first time that this data source has been analyzed in such detail to provide insight into what CCOs do while they are deployed.

one location, and training for CCOs. Participants in interviews and focus groups also highlighted the fact that many of the factors that increased the stress of deployed contracting personnel would not be affected by reachback, but could be affected by other policy changes. As a result, we reached conclusions not only about the potential of reachback to reduce deployments but also about the need for policy and procedural changes to address other causes of stress.

Reachback and Deployments

The potential to decrease deployments by using reachback for purchases in certain categories is substantial, but reducing them enough to affect the deploy-to-dwell time may be a challenge because of the number of joint billets involved. We recommend that the Air Force do the following:

1. Use the analysis presented in this document to refine the estimates for potential decreases. (See pp. 31–34.)
2. Examine in more detail the reachback experiences of the organizations discussed in this document to learn how to take advantage of other benefits of reachback, such as strategic purchasing and the concentration of expertise for large source selections. (See pp. 21–27.)

Approaches to Relieving Other Sources of Stress

To address other areas that contribute to the stress on deployed CCOs, we recommend that the Air Force do the following:

1. Facilitate support for better requirements and statements of work development by customers. This will help relieve CCOs of what they consider to be a time-consuming "extra" duty. (See pp. 40–44.)
2. Work to consolidate requirements to fewer contracts. Our data analysis shows that CCOs in JCC-I/A wrote more purchase orders than any other type of contract—most of them used only once. Consolidation will help reduce the CCO workload. (See p. 12.)
3. Revise its policies to allow the deployment of lower-grade contracting personnel and the deployment of other career fields (such as acquisition) to positions that are currently filled by contracting personnel but do not actually require someone with a warrant.[3] This would increase the size of the pool of deployable personnel and would have the added benefit of providing training for inexperienced contracting and acquisition personnel. (See pp. 46–47.)
4. Periodically review deployed personnel allocations to ensure that the number and skill mix at a given deployed location is appropriate. (See pp. 46–47.)

---

[3] The extent of an individual's authority to obligate funds and commit the government contractually is expressly defined in a "warrant" or other instrument of delegation, such as orders or certificates of appointment (U.S. Government Personnel Management Office, 1983).

5. Clarify the roles of other personnel in the contracting process (e.g., contracting officer's representatives, quality assurance evaluators) so that CCOs can focus on their primary duty of purchasing goods and services for the warfighter. (See pp. 39–45.)

Acknowledgments

We thank our sponsor, Roger S. Correll (SAF/AQC), and his deputy, Pamela Schwenke, for their support throughout this research. Scores of military and civilian personnel were willing to spend time with us and answer numerous questions about contingency contracting over the course of this project, and we are grateful for their assistance. The easiest way to thank these individuals by name is to list them alphabetically by the organization they were part of, and by the ranks they had, at the time of this research:

- **Air Force Contract Augmentation Program (AFCAP):** James Garred, Chief, Air Force Civil Engineer Support Agency Support Flight; Wayland Patterson, Division Chief/ AFCAP
- **Air Force Center for Engineering and the Environment (AFCEE):** Lt Col Steven Alpers, Chief of Contingency Contracting at the Air Force Center for Engineering and the Environment (AFCEE/AQC); Col Geoffrey Ellazar, head of AFCEE's acquisition and contracting division (AFCEE/AC); Michael Prazak and Craig Mayo of AFCEE's contingency construction division (AFCEE/CX)
- **U.S. Air Forces Central (USAFCENT):** Lt Col Jonathan J. Swall and Lt Col Andrew L. Sackett, USAFCENT/A7K (Contracting)
- **Hill Air Force Base (AFB):** Sharon Porter (contact for Hill AFB focus groups)
- **Joint Contracting Command Iraq/Afghanistan:** Col Roger H. Westermeyer, JCC-I/A/ J3, Principal Assistant Responsible for Contracting, Iraq; LTC James R. Bachinsky, JCC-I/A (Green Zone); Capt David L. Zimmerman, JCC-I (Camp Victory); Capt Jake Alverson, JCC-I (Camp Victory); Dean Carsello, JCC-I/A/J3, Stateside Liaison Officer
- **Randolph AFB:** Lt Col Rene Richardson, Commander, Air Education and Training Command Contracting Squadron (AETC CONS); SMSgt George Brown (contact for Randolph AFB focus groups)
- **Rock Island Contracting Center:** Mike Hutchison, Deputy Director, Rock Island Contracting Center, U.S. Army Contracting Command; Sue Phares, Chief, JCC/ Surface Deployment and Distribution Command (SDDC) Reachback Branch; Rock Woodstock, liaison to Afghanistan; Lt Col Michael Eastman, Principal Assistant Responsible for Contracting, Afghanistan
- **Office of the Assistant Secretary of the Air Force for Contracting:** Col Denean Machis, SAF/AQCX; Lt Col Nathan Rump, SAF/AQC; Lt Col Tara Morrison, SAF/AQCA
- **Shaw AFB:** Major James A. Hageman, Commander, 20th Contracting Squadron
- **Wright-Patterson AFB:** Col Thomas Snyder, HQ AFMC/PK (the contracting office of the Air Force Materiel Command); Capt Anthony F. D'Angelo, HQ AFMC/PK

Executive; Capt Derek J. Aufderheide, USAF AFMC ASC/PK (the Aeronautical Systems Center under AFMC) (contacts for focus groups).

We also extend our appreciation to the authors of the 236 AARs that we analyzed for this research. While we cannot list them by name, we would be remiss in not acknowledging the important contribution their accounts made to this research effort. Similarly, we appreciate the time and candor of our focus group participants, whose names are not listed to protect confidentiality. We also appreciate Col Westermeyer's help in providing us access to JCCS data.

We thank Elizabeth Wilke of RAND, who provided excellent note-taking support for our focus group discussions.

Finally, we thank our reviewers, Susan Gates and Raymond Conley of RAND and E. Cory Yoder of the Naval Postgraduate School, for their many excellent suggestions that helped improve the presentation of our results.

Abbreviations

20 CONS	20th Contracting Squadron
A&AS	advisory and assistance services
AAR	after action report
ACC	Army Contracting Command
ACO	administrative contracting officer
AEF	air and space expeditionary force
AETC	Air Education and Training Command
AFB	Air Force base
AFCAP	Air Force Contract Augmentation Program
AFCEE	Air Force Center for Engineering and the Environment
AFCESA	Air Force Civil Engineer Support Agency
AFMAN	Air Force manual
AFSC	Air Force specialty code
AMC	U.S. Army Materiel Command
AOR	area of responsibility
BPA	blanket purchase agreement
CCO	contingency contracting officer
CONUS	continental United States
COR	contracting officer's representative
CPA	Coalition Provisional Authority
DCMA	Defense Contract Management Agency
DFARS	Defense Federal Acquisition Regulation Supplement
DFAS	Defense Finance and Accounting Service

DoD	U.S. Department of Defense
ECONS	expeditionary contracting squadron
FAR	Federal Acquisition Regulation
FOB	forward operating base
FOO	field ordering officer
FPDS-NG	Federal Procurement Data System–Next Generation
FSS	Federal Supply Schedule
FY	fiscal year
GPC	government purchase card
GSA	General Services Administration
HERC	heavy engineering repair and construction
IDEAS	Interactive Demographic Analysis System
IDIQ	indefinite delivery, indefinite quantity
JARB	Joint Acquisition Requirements Board
JCC-I/A	Joint Contracting Command Iraq/Afghanistan
JCCS	Joint Contingency Contracting System
LOGCAP	Logistics Civil Augmentation Program
LSU	LOGCAP Support Unit
MAJCOM	major command
MNSTC-I	Multi-National Security Transition Command–Iraq
OEF	Operation Enduring Freedom
OIF	Operation Iraqi Freedom
PAF	RAND Project AIR FORCE
PCO	procuring contracting officer
PIIN	procurement item identification number
PM	program manager
POC	point of contact
QA	quality assurance
QAE	quality assurance evaluator
RCC	regional contracting center

RICC	Rock Island Contracting Center
RFP	request for proposal
RMS	readiness management support
SAF/AQC	Office of the Assistant Secretary of the Air Force for Contracting
SAP	simplified acquisition procedures
SAT	simplified acquisition threshold
SDDC	Surface Deployment and Distribution Command
SOW	statement of work
SSA	source selection authority
TWR	theater-wide requirements
UAE	United Arab Emirates
USAF	U.S. Air Force
USAFCENT	U.S. Air Forces Central
USCENTCOM	U.S. Central Command
WERC	Worldwide Environmental Restoration and Construction

CHAPTER ONE

Introduction

Background

Air Force personnel in the contracting career field purchase and provide supplies and services in a variety of environments. Contracting personnel act as business advisers, buyers, negotiators, and contract administrators for day-to-day operations at Air Force installations and perform contracting duties at headquarters offices and product centers that are responsible for the acquisition of major weapon systems.[1]

In fiscal year (FY) 2008, the Air Force had just under 800 officers in the contracting career field (Air Force Specialty Code [AFSC] 64P); 1,100 enlisted personnel (AFSC 6CXX) (Pianese, 2009c); and about 4,700 civilian personnel (General Service code GS-1102) (Pianese, 2009b) supporting some 71 organizations (SAF/AQC, 2008, slide 8). Another way of looking at the size of the contracting workforce is that the Air Force had about one military contracting person for every 173 military personnel.[2] In contrast, the Army had fewer than 400 military personnel in the contracting specialty in FY 2006, or about one for every 1,500 soldiers.[3]

During military operations or emergency situations that require the use of military forces, contracting personnel may be deployed (inside or outside the continental United States [CONUS]) as contingency contracting officers (CCOs) who can enter into contracts on behalf of the U.S. government and serve as business advisers to the deployed or on-scene commander (USAF, 2007, para. CC-102). When serving in this capacity, Air Force CCOs may provide support beyond the Air Force, to include supporting other services' installations and personnel, as well as filling joint billets.

This research was motivated by a concern of the Deputy Assistant Secretary of the Air Force for Contracting (SAF/AQC) about increasing stress that Air Force military contracting personnel are experiencing due to the deployment demands placed on them. Air Force contracting personnel deploy with Air Force units as part of the Air and Space Expeditionary Force (AEF) deployment cycle, but they also deploy to support other services as part of

---

[1] Air Force Manual (AFMAN) 36-2105, p. 348, describes officer contracting duties (USAF, 2004b); AFMAN 36-2108, p. 482, describes enlisted contracting duties (USAF, 2004a). The *Career Planning Guide for Contracting Professionals* (SAF/AQC, 2005) describes career progression and working environments for contracting professionals.

[2] In FY 2008, the Air Force had about 328,000 officers and enlisted personnel (Air Force Association, 2008, p. 48).

[3] In FY 2006, the Army had 279 officer, 62 enlisted, and about 5,500 civilian contracting specialists (Gansler, 2007, p. 93), for a military population of 502,790 (Maxfield, 2006). One of the recommendations of the Gansler Report was to increase the size of the Army's military contracting career field by 400 personnel.

joint organizations, such as the Joint Contracting Command Iraq/Afghanistan (JCC-I/A),[4] which has authority over all contracting activities (with the exception of the U.S. Army Corps of Engineers) assigned or attached to U.S. Central Command (USCENTCOM), and the Defense Contract Management Agency (DCMA).

Since FY 2001, deployment taskings for CCOs have increased by 350 percent, with most of the increase being the result of demands for personnel in JCC-I/A because of Operations Iraqi Freedom (OIF) and Enduring Freedom (OEF) (Machis, 2009). Figure 1.1 illustrates the impact of this change when compared to changes in the supply of military contracting personnel.

In Figure 1.1, the "Total deployed" line shows that from 2004 to 2008 the number of Air Force contracting personnel deployed as CCOs increased from 134 to 280. The figure also has a line labeled "Manning needed to maintain 1:2 dwell." This line shows the number of deployable contracting personnel the Air Force would need in order to allow them to stay home twice as long as they were deployed before deploying again. For example, in January 2004 when 134 CCOs were deployed, the Air Force needed to have 402 deployable military contracting personnel in order to maintain a 1:2 ratio (six months deployed and then twelve months at home, for example): While 134 CCOs are deployed, 134 would be at home getting ready to deploy and 134 would be at home having just returned from deployment.

As the "Total assigned" line in Figure 1.1 shows, from 2004 to 2008 the total number of military contracting personnel (AFSCs 64P and 6CXX) declined from 2,263 to 1,900. Captains and majors (officer ranks O3 and O4) and enlisted personnel in grades E5-E7 (staff

Figure 1.1
Changes in Military Personnel Assigned and the Number of CCOs Deployed

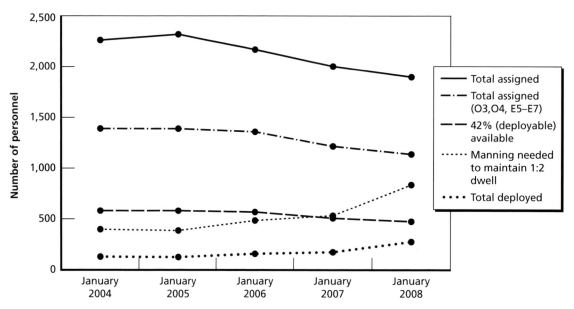

SOURCES: Total deployed data from Ray, 2007, and Rump, 2008. Total assigned data from the Air Force Personnel Center's Interactive Demographic Analysis System (IDEAS) database.
RAND TR862-1.1

---

[4] After this report was completed, CENTCOM established the CENTCOM Contracting Command, which is based in Qatar. JCC-I/A transitioned to the new organization in June 2010.

sergeant to master sergeant) constitute the majority of those who deploy; the inventory of personnel in these categories declined from 1,393 to 1,140.

Because of illness, people attending professional schools, and people filling positions that are designated "non-deployable," the actual pool of deployable personnel is significantly smaller than the number of personnel assigned. If we assume that only 42 percent of the O3s, O4s, and E5-E7s are truly deployable,[5] Figure 1.1 shows a problem arising in January 2007 when the "Manning needed to maintain 1:2 dwell" line crosses, and then rises above the "42% (deployable) available" line. This occurs when the number deployed (about 250) is one-third of the number available (about 750). After January 2007, the increase in deployment taskings meant that the career field could no longer maintain a 1:2 deploy-to-dwell ratio. In fact, in 2008 the Air Force decided to adjust deployments for the career field to a 1:1 ratio—i.e., six months deployed and six months home before deploying again (Holmes, 2008).

Other indications of stress on the contracting workforce are the heavy representation of Air Force personnel on the JCC-I/A staff (70 percent of this joint staff is Air Force), a decline in the average career length of 6C enlisted personnel from over 10 years in FY04 to under 6.5 years in FY 2008 (Pianese, 2009a) and a decline in the percentage of personnel at the "journeyman/craftsman" level of experience (11–15 years of service) in the career field from 20 percent in FY 2004 to 15 percent in FY 2008—a decline that means the career field will have fewer senior experienced personnel in the future (Machis, 2009).

SAF/AQC asked RAND to assess the feasibility of *reachback*, which we define as the use of contracting capability outside the theater of operations to accomplish contracting tasks for customers in-theater, as a potential means for alleviating stress on the Air Force's contracting workforce. Reachback is currently practiced by some organizations in the Air Force and the Army, and the perception exists that using it more could lead to fewer deployments (thus reducing stress) and even lower costs (e.g., for transportation and hazardous duty pay). Using reachback also might lead to increased contracting efficiencies and effectiveness because of continuity of the workforce (i.e., a reachback unit would not experience the regular changeover that deployed units do) and the concentration of highly skilled people in one place.[6]

Research Approach

We began this project accepting the assessment of SAF/AQC that the contracting career field is experiencing stress. We performed a detailed analysis of 236 CCO after action reports (AARs) from 1999 to 2008—all of the AARs that are maintained online by the Defense Acquisition University—to learn more about what CCOs actually do in-theater, what contracting-related problems they have encountered, and what might be amenable to reachback.[7] In summer 2009,

---

[5]   This was the actual percentage in October 2007 for AFSC 2T2 (air transportation and vehicle maintenance)—another career field in high demand. We use this value here because the actual number for 64Ps and 6CXXs was unavailable and our contacts in SAF/AQC indicated it was reasonable for illustrative purposes.

[6]   Note, however, that the Gansler Commission, which was established to address Army problems related to contingency contracting, was opposed to increasing the use of reachback. "Reachback to CONUS has not worked well due to the absence of timely interface with the warfighter" (Gansler, 2007, p. 6). The Gansler report provides no data to support this assertion.

[7]   Appendix B discusses what AARs are, our analytic approach to the AAR data, and descriptive statistics.

we also conducted a series of 11 focus groups (nine with CCOs who had recently returned from deployments and two with civilians and military personnel who had not deployed) to gain more insight into the causes of workforce stress and the potential use of reachback.[8] In addition to CCO focus groups, we conducted 12 interviews with other CCOs, personnel in management positions for contracting organizations, and members of organizations that use, or have used, reachback for contracting functions.[9] Finally, we analyzed Joint Contingency Contracting System (JCCS) data from FYs 2003 and 2008 on purchases for OIF and OEF and used those data to explore the potential impact on deployments of using reachback for certain categories of purchases.

As our research proceeded, we learned that, although CCOs recognized the potential of reachback to reduce deployments, they felt it had more potential to improve other aspects of contingency contracting. In addition, we learned of other factors that increased the stress of deployed contracting personnel that would not be affected by reachback but could be addressed by other means. Table 1.1 shows how the various data sources were useful in addressing our research questions.

Note that our data focus on CCOs; obtaining the customer perspective about the use of reachback was not part of the scope of this report. CCOs mentioned a number of concerns regarding customers' lack of knowledge of the contracting process, their failure at times to handle contracting-related responsibilities, such as base access and quality assurance, and their perceived need to have a CCO collocated with them. Accordingly, the implementation of a reachback-oriented solution may be more effective if customers' perspectives are solicited and taken into consideration.

Table 1.1
Research Questions and Sources Used to Address Them

Topic	AARs	JCCS Data	Interviews	Focus Groups
What CCOs do	✓	✓		✓
What is amenable to reachback	✓	✓	✓	✓
Potential reachback benefits	✓		✓	✓
Potential reachback downsides	✓		✓	✓
Sources of CCO stress	✓		✓	✓

Organization of This Report

The second chapter of this report uses evidence from AARs and JCCS data to develop a picture of what CCOs do when they are deployed. Chapter Three combines the experience of

---

[8] Appendix D includes a discussion of how these groups were selected.

[9] We contacted some interviewees because they managed organizations that use, or have used, reachback in some form. Other interviewees were recommended by our sponsor or by focus group participants. These individuals were considered to be subject-matter experts who had education, training, or experience related to contingency contracting operations, CONUS-based contract support for contingency operations, managing the Air Force's contracting workforce, and contracting data systems.

organizations that use reachback with insights from CCO focus groups to develop some basic criteria for the use of reachback. Chapter Four applies these criteria to JCCS data on contracting in OIF and OEF to make rough estimates of potential reductions in CCO deployments if reachback were employed for certain categories of purchases. Chapter Five continues the examination of focus group and interview data and discusses stresses on CCOs that might not be decreased by reachback but could be addressed by other means. That chapter also poses some questions related to the implementation of reachback. Finally, Chapter Six presents our conclusions and recommendations. Appendix A provides details about some organizations that use reachback (Air Force Contract Augmentation Program [AFCAP], Air Force Center for Engineering and the Environment [AFCEE], and the JCC Reachback Center). Appendix B outlines the approach used for the analysis of AARs. Appendix C describes JCCS data in more detail, and Appendix D includes the interview protocol used in the focus groups.

CHAPTER TWO

What Contingency Contracting Officers Do

In this chapter, we show how JCCS and other data can be used to better understand the nature of the work done by deployed CCOs. We then use data from AARs and focus groups to learn more about how that work is accomplished.

What CCOs Buy and How They Buy It

Spending in the Area of Responsibility

Total spending on purchased goods consumed and services performed in the USCENTCOM area of responsibility (AOR) in FY 2008 was almost $28 billion. Figure 2.1 shows that about $16 billion of that total was spent in Iraq, $6.5 billion in Afghanistan, and about $5.5 billion in Kuwait, the United Arab Emirates (UAE), and Qatar; Figure 2.2 shows the number of contracts and contract actions associated with these purchases. Most of this spending was on contracts primarily written by contracting offices located outside the AOR and recorded in the Federal Procurement Data System–Next Generation (FPDS-NG), as shown by the "FPDS only" part of the bars in Figure 2.1.[1]

CCOs with the JCC-I/A wrote contracts and contract actions for more than $7.5 billion of goods and services performed in the AOR to support USCENTCOM operations in FY 2008. Most of these purchases are recorded in the JCCS and FPDS-NG data systems, though some are only in the JCCS. In Figures 2.1 and 2.2, the "JCC-I/A and FPDS" and "JCC-I/A only" portions of the bars show purchases made by CCOs in-theater; most of that spending occurred in Iraq and Afghanistan. Even though JCC-I/A CCOs were responsible for only about 34 percent of the dollars spent on work performed in Iraq and Afghanistan, Figure 2.2 shows that they wrote about 84 percent of the contracts and 82 percent of the associated contract actions. Most of those contracts and contract actions are recorded in the JCCS data system. The high ratio of contracts and actions to dollars suggests a high prevalence of tactical buying (i.e., contracts written in response to an operational requirement) rather than using

---

[1] The data supporting Figures 2.1 and 2.2 are in Appendix C, Table C.1. Data sources for the figures are the JCCS and the FPDS-NG information systems. Federal Acquisition Regulations (FAR) require that all contract actions of value equal to or greater than $2,500 be recorded in FPDS-NG. Contingency contracts are exempted, though even most of those are now being recorded in FPDS-NG. Data in Figures 2.1 and 2.2 are for U.S. Department of Defense (DoD) organizations only. They do not include security contracts written by the Department of State or humanitarian assistance provided by the U.S. Agency for International Development. FY 2008 JCCS data were downloaded from the JCCS website on January 22, 2009, and FY 2008 FPDS-NG data were downloaded on April 18, 2009. Common contracts are in the category "JCC-I/A and FPDS"; contracts only in the JCC-I/A data are in the category "JCC-I/A only"; contracts only in FPDS are in the category "FPDS only."

contracts written prior to actual requirements based on anticipated requirements. By way of comparison, Figures 2.1 and 2.2 also show purchases made for Kuwait, Qatar, and the UAE.

Figure 2.1
FY 2008 Spending for the USCENTCOM AOR

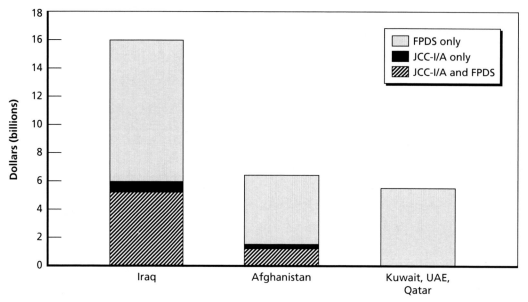

SOURCES: FY 2008 JCCS data and FY 2008 FPDS-NG data.
RAND *TR862-2.1*

Figure 2.2
FY 2008 Contracts and Contract Actions for the USCENTCOM AOR

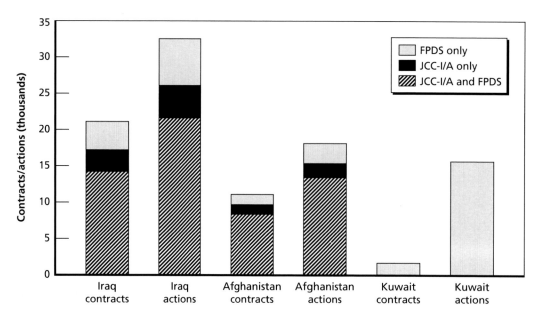

SOURCES: FY 2008 JCCS data and FY 2008 FPDS-NG data.
RAND *TR862-2.2*

Spending and the number of contract actions in those three locations were only slightly lower than in Afghanistan, but the number of contracts (which were all written by contracting offices located outside the AOR) is significantly smaller, suggesting that more buying at these locations was done using contracts written prior to actual requirements.

General Categories of Purchases

In the JCCS data, purchases are broken into three broad categories: services, commodities, and construction. Figure 2.3 provides the spending, contracts, and contract actions in these categories.[2]

Slightly more than half of the JCC-I/A spending by CCOs was for services (about $4.1 million), but over 70 percent of the contracts themselves and 63 percent of contract actions were for *commodities* (items that can range from protective barriers and gravel to light-bulbs and spare parts). Figure 2.3 also shows that about half of the contracts and contract actions for commodities went to local vendors whose headquarters were located in Iraq or Afghanistan, and more than half of the money spent on commodities was paid to local vendors. This suggests that efforts are being made to implement the "Iraqi first" and "Afghan first" policies in the AOR. These policies were put in place to purchase from local vendors in order to stimulate local economies, decrease unemployment, and engage local populations in activities conducive to establishing secure environments. However, while the majority of contracts and contract actions for services are with local vendors, only about a quarter of the spending on services has gone to local vendors.[3]

Contractual Instruments Used

The $7.5 billion spent by CCOs working for JCC-I/A in FY 2008 came through the award of 41,433 contract actions from 26,889 contracts of various types. These contract types include the following:

- **Purchase order.** Purchase orders are usually for one-time purchases with one delivery. The simplified acquisition threshold (SAT) below which a CCO can award a contract without seeking higher-level approval for contingencies outside CONUS is $1 million (FAR, 2009, Part 2.101).[4]
- **Blanket purchase agreement (BPA).** A BPA is an order made against an agreed-upon set of terms and conditions. It is not a contract but is a charge account set up for small purchases to be made on a repetitive basis with qualified sources and avoids the administrative need to write multiple purchase orders. BPAs require that funding documents be set up to provide the requisite financial resources.

---

[2] Detailed data behind Figure 2.3 are shown in Appendix C, Table C.2.

[3] There are far fewer service contracts than commodity contracts. Services provided by nonlocal vendors are complex, long term, and generally not available in the local economy (for example, providing security, creating a national maintenance depot, and providing purified bottled water services). Services provided by local vendors are more economical to buy locally, less complex, and targeted toward developing the local economy (for example, food, laundry, trucking/long haul, and vehicle maintenance).

[4] Simplified acquisition procedures are authorized for acquisitions that do not exceed $11 million when the acquisition is for commercial items that are to be used in support of a contingency operation (see FAR 13.500(e)(1), 2009). The simplified acquisition threshold for noncontingency purchases is $100,000. It goes up to $250,000 for contract awards or purchases made for contingencies within CONUS.

Figure 2.3
FY 2008 JCC-I/A Dollars Spent, Number of Contracts, and Actions, by Commodity Class and Vendor Type

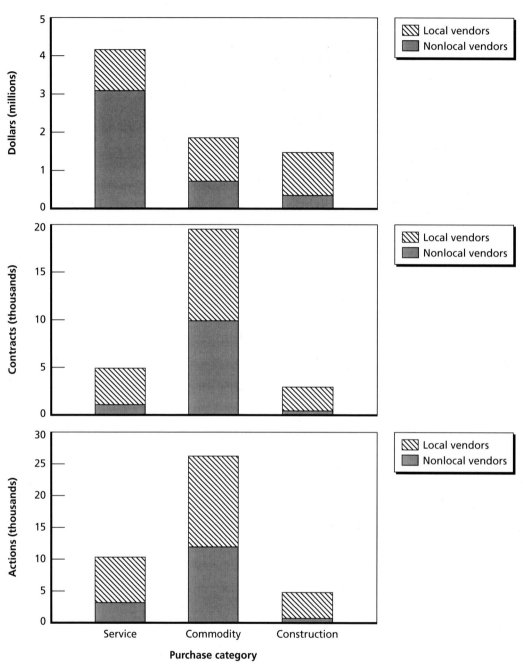

SOURCE: FY 2008 JCCS data.

RAND TR862-2.3

- **Contract.** This is a term used in the JCCS data. It refers to an indefinite-delivery, definite-quantity contract in which the requirement is known, such as a known quantity of items needed, but not the schedule or when the items will be ordered from the vendor.
- **IDIQ contract.** An indefinite-delivery, indefinite-quantity (IDIQ) contract is written to include a variety of requirements that might materialize. Specifics of the requirements and delivery or schedule are unknown at the time of the contract award. IDIQ contracts tend to be among the larger, more complex kinds of contracts.
- **FSS/GSA schedule.** The Federal Supply Schedule (FSS) is a listing of contractors who have been competitively awarded a General Services Administration (GSA) contract and can provide goods and services required by the contract. A GSA schedule is a collection of contracts listing the set prices for commercial goods and services that government organizations may use.
- **GPC.** A Government Purchase Card (GPC) can be used in lieu of a contract to purchase small dollar value goods and services called "micro-purchases" over the counter or on the open market or to make a payment for a purchase made from an existing contract. GPC micro-purchases are limited to $25,000 for contingencies outside CONUS (see the Defense Federal Acquisition Regulation Supplement [DFARS] Part 213.301).
- **SF44.** SF44 is a purchase order form, Purchase Order-Invoice-Voucher, for on-the-spot, over-the-counter purchases of goods and services available in the general market, to be used when away from a contracting office or in isolated areas. It is used to record a contract action at or below the micro-purchase level, i.e., $25,000 for contingencies outside CONUS.
- **Other.** Other miscellaneous contract types, such as leasing agreements, for amounts that are smaller than those in the categories listed above.

Table 2.1 lists the awards by contract type in decreasing order of the number of contract actions and provides several details related to each contract type.

Table 2.1
Types of Contingency Contracts, FY 2008

Type	Contracts		Actions		Dollars		Dollars per Contract (1000s)	Dollars per Action (1000s)
	Number	%	Number	%	Millions	%		
Purchase order	19,058	71	22,812	55	1,353	18	71	59
BPA	483	2	6,485	16	713	9	1,476	110
Contract	3,070	11	4,550	11	2,216	29	722	487
IDIQ contract	301	1	2,953	7	2,791	37	9,273	945
FSS/GSA schedule	1,594	6	1,912	5	385	5	242	201
GPC	1,666	6	1,754	4	4	<1	3	2
SF44	292	1	310	1	0.7	<1	2	2
Other	425	2	657	2	54	1	127	82
Total	26,889	100	41,433	100	7,517	100	279,571	181,434

SOURCE: FY 2008 JCCS data. Data downloaded from JCCS website on January 22, 2009.

For example, purchase orders made up 71 percent of the contracts that CCOs wrote in FY 2008, 55 percent of the actions, and 18 percent of the dollars. The size of purchase orders averaged $71,000 per contract and $59,000 per action. Most of the purchase orders in JCCS had only one action. A similar buying pattern for purchase orders exists in CONUS, although in CONUS more contracting officers are available to spread the workload. JCC-I/A recognized that certain common requirements were better off contracted at the theater-wide level and established an office of theater-wide requirements (TWR) to handle these kinds of contracts. Certain industrial sectors, such as electrical, water and power, and finances and security, were developed through theater-wide contracts of a contract or IDIQ type. These contracts are larger, cover more requirements, are of several years duration, and typically require a source selection authority as their dollar values exceed certain thresholds.

We also analyzed the JCCS data for the extent to which contracts were used by more than one contracting site. These results are displayed in Table 2.2. Many requirements are site-specific, but other particularly common requirements, such as gravel and barriers, are not. The use of fewer, larger contracts tends to reduce CCO workload because once they are set up, it takes minimal effort to order from them. We examined the extent of using contracts across contracting sites and found that (as shown in the first row of Table 2.2) most contracts (98 percent), contract actions (93 percent), and dollars (57 percent) were associated with only one contract site. As seen in the third row of the table, 64 contracts valued at $442 million were used by two contracting sites. Only a handful of contracts were used by three or more sites.

Table 2.2
Use of Contracts by More Than One Location

Number of Offices		Number of Contracts	Percentage of Contracts	Number of Actions	Percentage of Actions	Dollars (millions)	Percentage of Dollars
Writing contracts	Using Contracts						
1[a]	1	26,215	98	38,429	93	4,250	57
1[b]	Multiple	587	2	2,016	5	2,488	33
1	2	64	<1	553	1	442	6
1	3	6	<1	213	1	225	3
1	4	5	<1	66	<1	46	1
1	5	1	<1	26	<1	2	<1
1	7	2	<1	130	<1	64	1
Total		26,880	100	41,433	100	7,517	100

SOURCE: FY 2008 JCCS data. Data downloaded from JCCS website on January 22, 2009.

[a] Data in this row are for contracts with a single buying office and a single user location. Contracts written at these offices are used only by these locations.

[b] JCC-I/A also has TWR offices that purchase commodities, construction, and services from a single buying location for multiple using/customer locations. These kinds of purchases include the consolidated buying of electrical, oil, and water services. Data in this row are for TWR contracts.

CCO Perceptions of Their Work

According to the Federal Acquisition Regulation, *acquisition* means

> [T]he acquiring by contract with appropriated funds of supplies or services (including construction) by and for the use of the Federal Government through purchase or lease, whether the supplies or services are already in existence or must be created, developed, demonstrated, and evaluated. Acquisition begins at the point when agency needs are established and includes the description of requirements to satisfy agency needs, solicitation and selection of sources, award of contracts, contract financing, contract performance, contract administration, and those technical and management functions directly related to the process of fulfilling agency needs by contract. (FAR, 2009, Part 2.101)

Basic steps in the process are presolicitation planning (including market research and refining requirements), solicitation preparation, proposal evaluation, source selection, contract award, and contract administration.[5] In this section, we look at the phases of the process that CCOs mentioned most frequently in AARs and in focus group discussions.

While policy, doctrine, and other official publications are key to understanding a CCO's formal responsibilities while deployed, accounts from CCOs themselves about how they spent their time during deployment are also an important empirical source of information. Firsthand descriptions of their deployment experiences can complement—or potentially contradict—Air Force documentation. Moreover, CCOs' accounts of how they spent their time in-theater may provide a basis for determining whether manning documents (documents that quantify the manpower required to accomplish certain tasks) accurately reflect how much time is needed to perform contracting tasks and include the entire set of contracting-related functions that CCOs perform while deployed. Additionally, CCO descriptions of their work may reveal contracting duties that formally fall under their purview, such as database updates and contract closeout, yet are being neglected.

Data from both AARs and focus groups proved to be an excellent source of this type of information. Specifically, during our analysis of CCO AARs, we tagged passages that indicated how CCOs spent their time while deployed. Since there was neither an explicit question nor any other type of prompt that instructed CCOs to systematically summarize their tasks or primary responsibilities, not all the AARs contained such discussions. Further, among the AARs that did include some reference to duties, tasks, or responsibilities, the context in which they were mentioned varied. For instance, some AAR authors began their report with an overview of their responsibilities, whereas others referred to their tasks within the context of broader observations, such as contracting problems encountered or recommendations for leadership. Overall, we coded references to or descriptions of tasks in 217 of the 236 AARs in our dataset.[6]

This analysis of AAR data was complemented by a review of notes from the nine focus groups we conducted with recently returned CCOs (recall that the remaining two focus groups

---

[5] Different documents have variations in the steps, and the steps may vary with the contractual instrument used and with the value of the contract. This list is a compilation from the "File Compliance Checklist" in the *Joint Contingency Contracting Handbook* (DPAP, 2009), Air Force Manpower Standard 12A0 (USAF, 2001), and notes from the Defense Acquisition University courses "Source Selection: The Best-Value Process" (CON 111) and "Mission Focused Contracting" (CON 120).

[6] Appendix C describes what we mean by "coding" these references.

were made up of civilians and military CCOs who had yet to deploy). During the course of the focus groups, we expressly asked CCOs several task-focused questions (e.g., which tasks took the most time, which fell along the wayside when they were busy); accordingly, responses to those questions were a rich source of data. Additional focus group passages, including discussions of primary responsibilities, references to challenges faced during deployment, and observations related to reachback, also contained task-related comments.

After coding CCO observations of the tasks they performed in-theater, we mapped them to different aspects of the contracting process, starting with planning-related responsibilities and other presolicitation functions and ending with contract administration following an award. References to procurements and simplified acquisition procedures (SAP)[7] were assigned to their own category, as were remarks pertaining to contracting operations oversight. Figure 2.4 depicts the overall contracting process, along with the frequency with which AAR authors and focus group participants mentioned each category of tasks. For example, 37 percent of AARs and seven of the focus groups included references to tasks that we classified as part of the presolicitation and planning stage. At the other end of the process, 62 percent of the AARs and all nine of the focus groups mentioned tasks related to post-award contract administration. Note that these figures do not indicate how much time CCOs spent on a type of task or responsibility; rather, they are best considered as indicators of issues that, upon reflection about their deployment experience, they think other CCOs should be aware of. Overall, responsibilities related to post-contract award administration, contracting operations oversight, and contract awards were most frequently cited by CCOs in their accounts, while source

Figure 2.4
Frequency with Which CCOs Discussed Their Contracting Tasks, by Phase

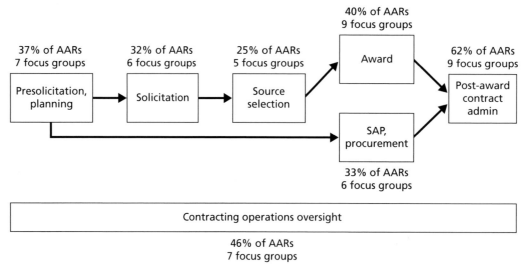

SOURCES: CCO AARs, 2009 RAND focus groups.
NOTE: N = 236 AARs, nine focus groups.
RAND *TR862-2.4*

---

[7] SAP are authorized streamlined purchasing methods described in FAR Part 13 that are used to expedite purchasing support to the warfighter. These simplified procedures are authorized for contracts with value less than a simplified acquisition threshold. See USAF, 2007, para. CC-102.

selection was mentioned less often. In the section that follows, we describe each category of contracting tasks and provide examples from AARs and focus groups that illustrate them.

Contracting Tasks Breakdown

Presolicitation Planning

The process begins with planning and presolicitation-oriented duties, including working with customers to develop a requirements package and/or statement of work (SOW), determining the most appropriate means of acquisition, and deciding on a suitable contractual instrument. As noted earlier, 37 percent of the AARs and seven focus groups featured discussion of these early-stage tasks. Pertinent comments included the following:

> The customer calls you up and needs life support, and fast. So you don't worry about the process. You give them life support, and then you spend the next week helping the customer write the statement of work for the help. You have to put together a requirements package. (Hill Air Force Base [AFB] Focus Group 2)[8]

> I also went to Ploce, Croatia, to conduct a pre-proposal site visit for the transfer of equipment from Ploce to Indonesia in support of their recovery efforts. (Bosnia AAR)

> My team worked hand in hand with customers to definitize requirements to ensure the items they wanted were actually what they were receiving. (Pakistan AAR)

> We didn't do much market research. We were supporting the national Afghani police, so we spent a lot of time putting proposals together. We worked projects with the Seabees [Navy construction unit], constructing guard towers, police headquarters, and border patrol checkpoints for different regions. I spent about 25 percent of my time in establishing requirements with the customer. . . . (Hill AFB Focus Group 1)

As these remarks suggest, interactions with customers to develop requirements predominated in this process category.

Once a decision regarding the means of acquisition has been made, the process continues onward either with a competitive approach, or, if the purchase is below the micro-purchase threshold, the CCO will go about directly acquiring the desired good or service.[9] SAP and procurement-related tasks include purchases made using a GPC, going into the local community to procure items, and acquiring items from retail outlets on base.[10] One-third of AARs

---

[8] After each focus group quotation, a unique identifier indicates the focus group in which the comment was made. Throughout this report, the same identifier is used to denote the same focus group but otherwise has no significance. These numerical identifiers are used to convey the extent to which evidence is present in multiple focus groups.

[9] The micro-purchase threshold for contingencies outside CONUS is $25,000; for these purchases, competitive bids are not necessary. For contracts for supplies or services in an overseas contingency, the simplified acquisition threshold is $1 million (FAR Part 13.003). Simplified acquisition procedures are also authorized for acquisitions that do not exceed $11 million when the acquisition is for commercial items used in support of a contingency operation (see FAR Part 13.500(e)(1)). If the dollar value is above $25,000, there is a requirement to solicit a "reasonable" number of sources (Long and Clements, 2007, p. 147).

[10] FAR Part 13.201(b) states that the GPC is the preferred method to purchase and pay for micro-purchases. It can be used to purchase items or services available on price lists or to make a payment for purchases from an existing contract.

and six of the nine focus groups included descriptions of these types of purchases, such as the following:

> Contracting Officers performed over 55 repetitive micro purchases that reduced the time and effort for more complex contracting actions. Only one of the units supported deployed personnel with GPC cards. Due to the shortage of GPC holders, a CCO had to go down town for every micro purchase made. (Australia AAR)

> During my deployment, the base was operating on 110/120 power. Most of the buying I was doing for CENTAF was done with GPC cards, to buy stateside what was difficult to get in country to support power in Baghdad. (Hill AFB Focus Group 2)

> In Afghanistan, about 75 percent of my time was taken up by micro-purchases. (Wright-Patterson AFB Focus Group 2)

> Duties consisted primarily of Base Operating Service (BOS) contract support, construction, Government Purchase Card (GPC) and local procurement of goods and services. (Iraq AAR)

Solicitation

If simplified acquisition procedures are not used, the next stage is solicitation. We coded references to tasks like drafting requests for proposals (RFPs) or bid requests and meeting with vendors to discuss project specifications as falling within this category. Exemplary comments included the following:

> As no one besides the contracting officers had experience in preparing solicitations and award documents, we required the governorate[11] project officers to submit the technical packages (i.e. bid schedules, delivery schedules, quantities, specifications, drawings, pictures etc) to us. We in turn then prepared the solicitations to include the evaluation criteria, terms and conditions prior to the release of the solicitation by the governorate project officers to the vendor base. (Iraq AAR)

> I had to go off base a lot to meet with vendors. That took a lot of my time. If you met with them face to face, that brought the price down a lot. You could name your own price, almost. You can say to the guy, "That's too high, I know it's too high," and get them to lower the price. Other people couldn't do that. (Randolph AFB Focus Group 1)

> The contracting team worked diligently to bring in new contractors to support the exercise. A vendor's conference was held to assist in this endeavor. (Thailand AAR)

About one-third of the AARs and six of the focus groups were coded for references to solicitation-oriented duties.

---

[11] Iraq is divided into 18 governorates, or provinces.

Source Selection

The contracting process moves from solicitation to source selection. When classifying tasks during our analysis, we considered source selection to include all responsibilities related to evaluating the merit of vendor proposals and making a decision on the award. In other words, this category includes formal or strategic source selections as well as other processes that CCOs engaged in to determine the most suitable vendor. Interactions with the customer to reach an award decision were also captured in this phase. The remarks that follow help illustrate this stage of the contracting process:

> At Manas Air Base [Kyrgyzstan], we contracted for services—we did more than ten source selections. We wrote three-year contracts instead of five-year contracts. (Shaw AFB Focus Group 1)

> As the most experienced CCO, with 17 years in contracting and the most source selection experience, I completed all the large dollar service contracts requiring source selections during my rotation. (Afghanistan AAR)

> A lot of what I did was major source selections, something [like] millions or billions of dollars. I didn't learn that doing operational contracting. (Wright-Patterson AFB Focus Group 4)

> CA [Civil Affairs] would frequently submit requirement packages with an insufficient sole source memo with very little vendor information. In their attempt to pull the local Iraqi tribal/Government leaders into more decision-making roles, they would let the local Iraqi Council make the decision as to which local tradesman would get which project. I put these projects out for bid to the suggested source and a few of my proven responsible contractors and make the decision based upon the outcome of the RFQ [request for qualifications] and my assessment of the contractor's capability. (Iraq AAR)

Award

After a vendor has been selected, a contract is awarded. Accordingly, we grouped tasks related to drawing up the appropriate contracting documentation as well as more-generic references to buying or awards in the award phase of the process. Forty percent of AARs and all nine focus groups contained references such as these award-focused remarks:

> We are strictly there to buy, and that's what we did. Eighty percent of our time was spent buying. Only a CCO could do that. (Hill AFB Focus Group 1)

> I processed 33 contract actions with a cumulative contract value in excess of $60 million. (Iraq AAR)

> I managed a multi-million dollar budget, administered six Blanket Purchase Agreements, procured 290 line items and awarded three construction contracts during my deployment. (Jordan AAR)

> . . . BPAs were used for cell phones and transportation requirements. Also, I established BPAs for water, ice, and fleet service. (Thailand AAR)

Although the percentage of AARs with references to awards may seem low given that awarding contracts is a primary formal CCO duty (USAF, 2001, Attachment 5), recall that CCOs are not instructed to list all tasks or provide time estimates for their contracting duties; they are asked to address problem areas and provide information that will be useful for the people who come after them. Accordingly, the tasks described in AARs are those considered important by the authors of those documents.

Post-Award Contract Administration

AAR and focus group data suggest that CCOs in-theater attended to many contract administration tasks following an award: Sixty-two percent of AARs and nine focus groups featured remarks about such contract administration.[12] General post-award contract administration comments included the following:

> We get involved in everything from beginning to end. Our job is to be business advisors; we just get roped into everything else, like logistics. A big issue for me was getting things through customs onto base. At Qatar, we had CCOs going to the gate to pick things up! (Wright-Patterson AFB Focus Group 3)

> Heavy equipment leases continue to be a large administrative effort. Setting up maintenance yards, access to base for contractors, and extremely high customer expectations add to challenging contract administration. (Iraq AAR)

> The majority of our contracts were administered by the Defense Contract Management Agency (DCMA) and suffered due to the lack of experience. We spent more time telling and showing them how to do their job when it would have been quicker to do it ourselves. (Iraq AAR)

As we coded the AARs and focus group notes, we observed a large range of discrete tasks that fell under the broad category of post-contract administration. Hence, we refined this category further, parsing out such tasks as base access and escort arrangements for vendors, delivery coordination, vendor payment issues, performance monitoring, and other more-traditional CCO tasks like updating databases and contract closeout. To illustrate, Table 2.3 features examples of some of the most often cited tasks that we regarded as post-contract administration.

Contracting Operations Oversight

The final category of tasks we examined pertains to oversight of processes and responsibilities that support contracting operations. As Figure 2.4 suggests, these tasks span the entire contracting process and include educating personnel about the contracting processes and

---

[12] In-CONUS post-award responsibilities can be divided among several people and organizations. The contracting officer manages the contract and vendor, including awarding and executing contract actions and managing schedules. If assigned by the procuring contracting officer (PCO), DCMA performs oversight of contract activities at the place of performance and can modify contract actions at the request of the contracting officer (DCMA, 2009, para. 7.4). The customer's contracting officer's representative (COR) monitors whether the vendor is providing goods or services in accordance with the contract's requirements, and the quality assurance representative monitors whether what the vendor is providing is of acceptable quality (see DFARS Part 201.602-2, "COR Responsibilities"; and FAR Part 46.4, "Government Contract Quality Assurance"). According to AAR and focus group data, the divisions among these roles became blurred, with the CCO assuming responsibilities that went beyond what is normally required of the contracting officer in CONUS.

Table 2.3
Examples of Post-Award Contract Administration Remarks

Task	Remarks
Base access, escort arrangements	My focus was on passes and getting people on base. (Shaw AFB Focus Group 1)
	Sometimes the customers did not have the ability to go out to the gate and escort the contractor, therefore one of the contracting officers was required to escort the contractor to avoid claims for the additional wait outside of the gate. (Afghanistan AAR)
Delivery coordination	Customs gets tied up. Sometimes it takes 10 to 20 days to get the customs waiver. There's no rep for us at customs in Qatar. (Wright-Patterson AFB Focus Group 1)
	The vendors don't typically share with us the fact that they aren't able to meet a deadline for delivery, which causes a lot of jumping through hoops and last minute phone calls to make sure that the mission is being met. (Iraq AAR)
Vendor payment	We do quality assurance, writing contracts, making sure people are getting paid. We do all that. (Wright-Patterson AFB Focus Group 3)
	The CCO makes weekly payments with Finance (ECPTS), at ECHO 1 gate to vendors who have delivered items to the base. (Kyrgyzstan AAR)
Performance monitoring	We needed to do surveillance of contractors if the work statement or contract was performance-based. We would do the quality assurance (QA) training. We had to follow up with the QAs and CORs to make sure they were doing what they're supposed to be doing. We were doing checks of QA, CORs instead of other tasks. (Shaw AFB Focus Group 1)
	I left the IZ (International Zone) several times to monitor two contracts and for the semiannual JCC-I/A conference. (Iraq AAR)
Other contract administration	We cut up modifications, and we update the contracts. (Randolph AFB Focus Group 1)
	Each buyer updated requirements in several logs to include the CENTAF tool . . . AEF log and their personal GPC log. (Iraq AAR)

SOURCES: CCO After Action Reports; 2009 RAND focus groups.

procedures, managing the professional development of contracting personnel, administering the field ordering officer (FOO) program, interacting with higher levels of military and civilian leadership, and furnishing data in response to audits and other external requests. The bulk of discussions concerning contract operations oversight tasks addressed either training and educating personnel (e.g., vendors, customers, commanders, CORs,[13] other CCOs) or administering the FOO program. Examples of education-oriented remarks include the following:

> On a daily basis though, we trained the vendors on how we did business and what our expectations were. (Afghanistan AAR)

> Lots of customers wanted to go directly to LOGCAP [Logistics Civil Augmentation Program].[14] I did a lot of educating the customer on how contracting worked and how they would have to go through the process, either to LOGCAP or go local. There were also a lot of subcontracts and layers to work out. Lots of people bypassed the system and went

---

[13] CORs are qualified individuals appointed by the CCO to assist in the technical monitoring or administration of a contract (Long and Clements, 2007, p. 198).

[14] LOGCAP is briefly described in Chapter Three.

to LOGCAP directly. I had to educate people about the process. (Wright-Patterson AFB Focus Group 2)

To better assist customers in the development of their requirements, we provided customers with a Customer Guide, a Guide on Writing SOWs (Statements of Work), templates of quality SOWs, and we offered to assist them in the process at every step of the way. (Kosovo AAR)

The frequency of customer turnover requires good documentation and constant training of resource managers, ordering officers, and Contracting Officer's Representatives (CORs). (Iraq AAR)

The FOO program, which typically involved Army personnel who are forward-deployed and authorized to make small, local purchases that are mission-critical, required Air Force CCOs to monitor FOO purchasing activity. This entailed "clearing" FOO transactions on a regular basis and ensuring that FOOs adhered to contracting regulations in their conduct.[15] Comments about the FOO program tended to resemble these excerpts:

The whole country has hundreds of field ordering offices. We cleared all of Afghanistan's FOOs, but they had to come to Kabul to be cleared. Later, we had a team that went on site out to their locations to clear. (Shaw AFB Focus Group 1)

Moreover, I was responsible for over 500 Field Ordering Officers who made cash purchases to support combat soldiers in the field. (Iraq AAR)

Overall, contacting operations oversight tasks were cited in 46 percent of the AARs and seven focus groups.

In Chapter Five, we use AAR and focus group data to describe problems that CCOs experienced and duties that they felt detracted from their primary responsibility—namely, contracting for goods, services, and construction.

Conclusion

Available data sources provide a significant amount of information about the tasks that CCOs accomplish and the purchases they make. JCCS data suggest that CCOs mostly conduct tactical buys (indicated by the high ratio of contract actions to dollars spent) for contracts purchasing goods or services for only one location. Comments indicate that CCOs "get involved in everything from beginning to end" (Wright-Patterson AFB Focus Group 3) in the purchasing process—sometimes in areas that may detract from their primary duties. As we show in the chapters that follow, insights from these data sources helped guide our exploration of the potential benefits of increasing the use of reachback for some contracting functions and workload.

---

[15] FOOs are not warranted and require contracting officers to reconcile or "clear" transactions after the fact. Transactions are limited to $2,500.

CHAPTER THREE

Reachback Potential

The previous chapter illustrated, both through CCO comments and contracting data, the nature of the work performed by CCOs in various theaters of operation. Equipped with detailed information about what CCOs do in-theater, as described by the CCOs themselves and as revealed in JCCS data, we can begin to explore the possibility of using reachback to accomplish some of these tasks.

In this chapter, we review the experience of organizations that currently use (or have used in the past) reachback for certain contracting functions, as well as data from focus groups and interviews, to develop criteria for what is and is not amenable to reachback.

Organizations Used for Reachback

Interviews with CCOs and managers in Air Force contracting offices helped us identify organizations that had experience with using reachback for different contracting functions. Studying how these organizations were structured and how they performed contracting tasks will help us understand the circumstances under which reachback can be used. We discuss these organizations briefly in the following order, roughly from those that use reachback for a wide variety of activities to those that use it for more-targeted functions:

- The Joint Contracting Command Reachback Branch at the Army's Rock Island Contracting Center (RICC) in Illinois[1]
- The 20th Contracting Squadron (20 CONS) at Shaw AFB, South Carolina
- The Air Force Contract Augmentation Program (AFCAP), based at Tyndall AFB, Florida
- The JCC-I/A stateside liaison office in San Antonio, Texas
- The Air Force Center for Engineering and the Environment (AFCEE) located at Brooks City-Base in San Antonio, Texas.

Since some CCOs mentioned their interaction with the Army's LOGCAP in AARs and focus groups, we also provide a brief description of that organization.

JCC Reachback Branch, Rock Island Contracting Center
Army contracting reachback efforts began in September 2007 as a response to fraud and contract quality issues experienced in Kuwait. These problems arose from the fact that a massive

---

[1]   See Appendix A for more details about the structure of the Reachback Branch.

increase in workload and in the complexity of contracts was not accompanied by increases in staffing levels or the skill levels of Army contracting personnel.[2] A stand-alone reachback branch was established at RICC for the sole purpose of providing support to the Army's 408th Combat Support Brigade in Kuwait. Positive experiences with this reachback organization led the leadership of the Army Materiel Command (AMC) to discuss with JCC-I/A leadership the possibility of expanding reachback capability to contracting required by JCC-I/A.

A memorandum of agreement was signed in January 2009 that established a "CONUS-based JCC Reachback Branch," the purpose of which was to "reduce the JCC-I/A workload, provide personnel continuity and/or unique capabilities, and assist in executing strategic sourcing candidates" (RICC, 2009).[3] The JCC Reachback Branch provides "cradle-to-grave" contract services: pre-award policy and legal reviews, pricing, solicitation preparation and issuance, proposal evaluation, negotiations, contract preparation, contract award, post-award contract administration, and closeout support. It uses the web-based Army Single Face to Industry system to advertise contracts, which means that competition is open to bidders worldwide.

By March 2009, the JCC Reachback Branch had 12 people in its office at RICC in Illinois, and the office was growing. In addition, a military liaison officer from RICC was sent to JCC-I/A headquarters, and a civilian representative was sent to Afghanistan. The liaisons work with JCC-I/A to determine which contracts are good candidates for the RICC reachback organization. As of April 2009, the reachback branch was working with close to $2.27 billion in contracts for JCC-I/A, an approximate distribution of which is shown in Figure 3.1.

Note that advisory and assistance services (A&AS), acquisition support services, and miscellaneous services account for about 23 percent of the total. These are types of contracts that, as we will see later in this chapter, CCOs felt were generally amenable to reachback.

20th Contracting Squadron, Shaw AFB

According to focus groups at Shaw AFB, 20 CONS began acting as a reachback organization for U.S. Air Forces Central (USAFCENT) sometime during 2004 (Shaw AFB Focus Group 2). As contracting needs related to OIF increased, several individuals in 20 CONS who had associations with, or had previously worked with, USAFCENT headquarters started to ask whether 20 CONS could provide support to help with the logistics tail. Over time, support expanded from OIF to OEF, with contracts being managed at Shaw AFB instead of in Qatar, where the forward headquarters for USAFCENT is located.[4] Examples were postal services, communications, and a support contract for Manas Air Base in Kyrgyzstan that employed 40 contractors. Headquarters staff at Shaw AFB made the decision to do some theater-wide sup-

---

[2]   Mike Hutchison, Deputy Director, Rock Island Contracting Center, "Rock Island Contracting Center Reach-Back Support to Joint Contracting Command," briefing, January 12, 2009.

[3]   According to Johnson, 2005, strategic sourcing "is the collaborative and structured process of critically analyzing an organization's spending and using this information to make business decisions about acquiring commodities and services more effectively and efficiently."

[4]   We note that 20 CONS would normally not be expected to perform a reachback function; its purpose is to support the fighter wing at Shaw AFB. In-theater contracting support is normally provided by expeditionary contracting squadrons (ECONS).

Figure 3.1
Reachback Contracts in RICC

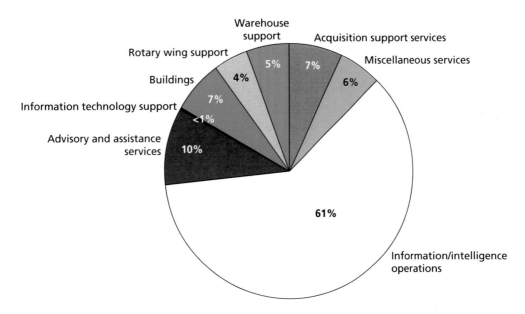

SOURCE: Author compilation from Westermeyer, 2009b.
RAND TR862-3.1

port contracts using 20 CONS. Like the later-established reachback cell at RICC, 20 CONS performed cradle-to-grave contracting services; unlike the RICC, however, contracts were open to U.S. bidders alone (Shaw AFB Focus Group 3).

Reachback support to USAFCENT put a significant strain on the squadron: Funds obligated by the organization more than tripled, from about $44 million in FY 2003 before it provided reachback support to $136 million in FY 2004 in the first year of that support. By FY 2008, 20 CONS had obligated $167 million, 80 percent of which was for USAFCENT (Hageman, 2009). Although the squadron established a reachback cell of four to five people who worked only for USAFCENT and not for the 20th Fighter Wing at Shaw AFB, these personnel were "taken out of hide"—there was no increase in manning for the organization despite the increase in work. Squadron leadership ultimately decided that the reachback demands were stressing contracting personnel and their ability to support the wing adequately, and reachback was terminated in October 2008.[5]

Air Force Contract Augmentation Program

AFCAP is designed for "rapid design/construction, service contracts and logistics/commodity solutions to mission requirements that arise in contingencies." Depending on the "urgency, degree of requirements definition or task stability, contracts can be tailored into firm-fixed price, cost-plus fixed fee or cost-plus award fee task orders" (AFCESA, 2010).

---

[5] As of July 2009, requests for reachback assistance are decided on a case-by-case basis with the exception of service contracts already in place. Some commodities ($500,000 and below) are being purchased, but no new service contracts are being written (Shaw AFB Focus Group 3).

AFCAP is a reachback organization in the sense that the contract instrument is managed by contracting officers in CONUS. Under AFCAP, five contractors compete for individual task orders that are placed by customers. In February 2009, AFCAP had six program management positions in-theater. DCMA provides administrative contracting officers (ACOs), but customers who use AFCAP are responsible for providing on-site contract administration, quality assurance, and task order surveillance. AFCESA AFCAP program managers "work as an interface between customers and the AFCAP contracting officers to solidify requirements available under this contract" (AFCESA, 2010).[6]

AFCAP can provide installation support services (such as base operating support), "contingent logistics" (e.g., vehicles and heavy equipment, humanitarian relief), and expeditionary construction (such as base camps and border fortifications) (Fryer, undated).

JCC-I/A Stateside Liaison Office

The JCC-I/A Stateside Liaison Office in San Antonio, Texas, reports to the deputy director of the operations directorate (J3) of JCC-I/A. This organization serves a limited but important reachback function by managing the closeout of expired contracts that were causing a backlog for CCOs in-theater. When JCC-I/A was established in 2004, it "inherited" coalition contracts from the Army Corps of Engineers, which had itself inherited contracts from the Coalition Provisional Authority (CPA). Some of the contracts on which the office is working expired in 2004, 2005, and 2006 but had not yet been closed out as of 2009.[7]

Air Force Center for Engineering and the Environment

AFCEE "provides integrated engineering and environmental management, execution, and technical services that optimize Air Force and Joint capabilities through sustainable installations" (AFCEE, undated).[8]

AFCEE's primary role in Iraq is to provide construction support for the Multi-National Security Transition Command–Iraq (MNSTC-I); 98 percent of its business in Iraq is for multinational forces.[9] This construction is done primarily through the Heavy Engineering Repair and Construction (HERC) contract, which uses 20 prime U.S. contractors and has a ceiling of $15 billion. AFCEE construction projects in Iraq include schools, medical facilities, ministry buildings, airports, and police stations (Fryer, undated).

AFCEE operates in Iraq and Afghanistan under a memorandum of understanding with JCC-I/A. None of the AFCEE personnel in Iraq are warranted; contracting officers with warrants are back in CONUS (and receive their warrants through the Air Force Materiel Command).

AFCEE has a forward presence in Iraq but not a forward contracting presence—a representative works with the MNSTC-I logistics staff, and there are eight or nine civilian and contractor project managers in-theater. By the time requirements are passed to AFCEE, they have

---

[6] See Appendix A for more details about AFCAP and how it functions.

[7] Interview with JCC-I/A Stateside Liaison Office, February 23, 2009. Once the last delivery has been made and the vendor has been paid, contract closeout requires a final reconciliation and documentation to ensure that contract terms and conditions have been met. DD Form 1597, *Contract Closeout Check-List*, describes what should be considered.

[8] See Appendix A for more details about AFCEE.

[9] Interview with AFCEE personnel, June 7, 2007.

already been vetted through several MNSTC-I staff organizations. A project manager sends the requirement to AFCEE, which puts together the requirements package with an RFP and sends it out to the 20 HERC contractors for proposals.[10]

The decision is made in the Gulf region as to whether the requirement is valid and, if so, whether it goes to AFCEE or the Army Corps of Engineers Gulf Region. All contracting work is done from Brooks City-Base, and AFCEE contracting officers do not deploy to Iraq or anywhere else.

The AFCEE contracts were not originally considered to be a component of *contingency* contracting capability, nor are they focused on military readiness. AFCEE got involved in OIF in 2003 when the CPA sought to save time in getting reconstruction projects started by using existing IDIQ construction contracts (SIGIR, 2006, p. 51). It was envisioned that AFCEE contract capability would help initiate commercial projects while the Army and the Corps of Engineers established an Iraq presence.

LOGCAP

LOGCAP is a way to "preplan for contingencies and to leverage existing civilian resources" in order to provide logistics and life support services to fighting forces (L. Thompson, 2009, p. 2). It was established by the Army in 1985 and has gone through several contract stages. The latest contract, LOGCAP IV, was awarded in 2007 to one contractor for program management and three other contractors for task order contracts (Parsons, 2009). Administratively, LOGCAP is under the Army Sustainment Command, which is subordinate to the AMC, and the executive director of LOGCAP is based at the Rock Island Arsenal in Illinois (Long and Clements, 2007, p. 141).

From a peacetime value in FY 2001 of $55 million, the contract grew to an annual value of $5.7 billion in FY 2008 (L. Thompson, 2009). Most of the work related to satisfying needs through LOGCAP is done in-theater: A deployed LOGCAP Support Unit (LSU) assists in drafting statements of work, a deployed ACO (usually provided by DCMA) obtains the cost estimate for the work, and a Joint Acquisition Requirements Board (JARB) in-theater validates the requirement. Upon funding approval, the PCO, located at the RICC, issues a notice to proceed with the work.[11]

Comparison of Reachback Organizations

Table 3.1 summarizes some of the characteristics of these organizations. For example, the "What Is Done" column shows that the JCC Reachback branch can work with a wide variety of strategic sourcing acquisition types and acquisitions that benefit from personnel continuity. 20 CONS, at the height of its reachback work, was instead more focused on commodities and services. AFCAP provides installation support services and expeditionary construction; AFCEE is exclusively for contingency construction. The JCC-I/A Stateside Liaison Office is used only for contract closeout. The "Who Does What" column shows that any warranted contracting officers related to these organizations are based in CONUS.

---

[10] Information in this paragraph is from an interview conducted at AFCEE on June 7, 2007.

[11] Information in this paragraph is derived from LeDoux, 2005, and focus group discussions.

Table 3.1
Organizations That Use, or Have Used, Reachback

Organization	What Is Done	Who Does What	Who Uses It	Work Accomplished
JCC Reachback Branch	Strategic sourcing, acquisitions that benefit from personnel continuity	Stand-alone JCC-I/A branch in CONUS does cradle-to-grave work Liaisons at JCC-I/A and in Afghanistan	JCC-I/A determines which contracts to send back to RICC	Currently working on close to $2.27B in contracts for JCC-I/A
20 CONS	Commodities (e.g., computers) and services (e.g., network support)	Squadron does cradle-to-grave work on contract Chief of QA at USAFCENT	ECONS in Qatar and Kuwait	$133M for 9AF/USAFCENT
AFCAP	Installation support services, contingency logistics, expeditionary construction	All contracting officers in CONUS 5 prime contractors DCMA provides ACO 6 PMs in-theater Customer provides QAEs and tech reps	". . . any DoD or federal entity"	AFCAP III: $10B ceiling (since November 2005, $601M, 126 task orders)
JCC-I/A Stateside Liaison Office	Contract closeout	Small team in CONUS works on closeout issues	JCC-I/A	"Leftover" contracts from 2004, 2005, and 2006
AFCEE	Contingency construction (e.g., Camp Bastion in Afghanistan)	All COs in CONUS 20 prime contractors About 10 PMs in-theater	MNSTC-I	HERC ceiling is $15B
LOGCAP	Logistics services and minor construction support for any scale operation	In-theater LSU manages planning and execution DCMA provides ACO DCMA may provide QA Units may provide COR	Primarily Army, but LOGCAP can support other U.S. military services as well as coalition and multinational forces	$5.7B in FY 2008

SOURCES: Notes from 2009 RAND interviews with these organizations and briefings provided by them. For LOGCAP, Long and Clements, 2007; Parsons, 2009; and L. Thompson, 2009.

NOTES: PM = program manager; QAE = quality assurance evaluator.

Table 3.2 highlights differences in how some of these reachback organizations provide contracting services. (It does not include all the organizations listed in Table 3.1; because the JCC-I/A Stateside Liaison Office has a limited mission and AFCEE's customers are more restricted than those of the other organizations, they are excluded.) For example, the processes for the JCC Reachback Branch and 20 CONS are essentially the same: Requests originate in-theater and the customer provides someone to monitor the work (either a QAE or COR), but everything else related to the contract is accomplished in the United States. As noted earlier, one difference is that, unlike 20 CONS, the JCC Reachback Branch can solicit bids from non–U.S.-based companies. AFCAP and LOGCAP are similar in purpose, but LOGCAP is set up to allow far more administrative work related to the contract to be accomplished in the theater of operations.

Table 3.2
Different Approaches to Reachback

Organization	Tasks Accomplished In-Theater	Tasks Accomplished in CONUS
JCC Reachback Branch	Request is developed by customer in consultation with RICC reachback representative at JCC-I/A	Policy and legal reviews Pricing, solicitation preparation and issuance (Army Single Face to Industry), negotiations, contract preparation and award, contract closeout
	Customer provides COR	*Worldwide bidders*
20 CONS	Request initiation through USAFCENT	Policy and legal reviews Pricing, solicitation preparation and issuance, negotiations, contract preparation and award, contract closeout
	Customer provides on-site QA	*U.S. companies only*
AFCAP	Request made through major command (MAJCOM) or government agency Bid evaluation with AFCESA (rare) Customer provides on-site QA DCMA provides ACO	Proposal preparation, bid evaluation, source selection, document preparation, task order issuance
LOGCAP	Request made through LSU LOGCAP deputy program manager advises combatant command JARB validates requirement ACO can approve new work	PCO at Rock Island can issue notice to proceed for new work

SOURCES: Notes from 2009 RAND interviews with these organizations and briefings provided by them. For LOGCAP, Long and Clements, 2007, and LeDoux, 2005.

The examples in Table 3.1 show that reachback has been used for a variety of process-related tasks, from very limited ones, such as contract closeout, to cradle-to-grave management of large service contracts. It is also used for different types of purchases, from advisory and assistance services to building air bases. Table 3.2 shows that different organizations have used different approaches to implementing reachback. We used these examples of current applications of reachback to guide some of the questions we asked in interviews and focus groups in order to gain more insight into the processes and requirements that might be amenable to reachback.

CCO Insight into What Is and Is Not Amenable to Reachback

Table 3.3 summarizes CCO insight into what considerations make a contracting task or process amenable to reachback. We divided these considerations into five categories: the requirement, the contract instrument, the size of the contract, the contract process, and other considerations. These categories and entries in the table are by no means all-inclusive; they represent the most common topics raised in interviews and focus groups.

Table 3.3
CCO Insight into What Is Amenable to Reachback

Requirement	Contract Instrument	Contract Size	Process	Other Considerations
Commodities	*GPC*	**Larger than $1M (above SAT)**	**Contract review**	*Theater-wide purchases*
Highly technical items	**GSA**		**Legal review**	
A&AS	Large IDIQ	**Larger than small/micro-purchase limit**	**Payment processing**	*Contracts that benefit from management continuity (e.g., long-term contracts)*
"Program"-type purchases (e.g., information technology)			Closeout	
			Formal or complex source selection	**Strategic sourcing**
Large construction projects				**Sufficient/reliable communication**
Services (trash, custodial)				**Accountability to local commander**
				Hours responsive to in- and out-of-theater needs

SOURCES: CCO AARs, 2009 RAND focus groups, and interviews.

NOTES: Standard font indicates thin or mixed evidence for the attribute being amenable to reachback. **Bold font** denotes attributes mentioned frequently or in more than one forum. ***Bold and italicized*** formatting together represent attributes with the strongest support—multiple sources and near-universal agreement.

In the "Requirement" column, *commodities* is rendered in bold and italicized text. This indicates that the recommendation that commodities (office supplies, computers) could be purchased using reachback and without the presence of a CCO in-theater was made in all of our data sources—AARs, focus groups, and interviews. A typical focus group comment was "All commodities can be reached back. The only reason it should be local is if [a commodity from] CONUS took too long to be shipped and was needed ASAP" (Randolph AFB Focus Group 1). The purchase of highly technical items (certain types of specialized equipment) also received near-universal support as a candidate for reachback. Other items in the "Requirement" column were mentioned frequently or in more than one forum. Note that A&AS is one of these; as noted in the discussion of the JCC-I/A Reachback Branch at RICC, reachback is already being used for some JCC-I/A A&AS contracts. Trash and custodial services were frequently mentioned as something that could be contracted using reachback, but they are listed in standard font because there was at least one comment that "smaller services" were best accomplished in-theater (Wright-Patterson AFB Focus Group 2).

In the "Contract Instrument" column, we see universal acknowledgment that purchases that could be accomplished with a GPC did not require the presence of a CCO in-theater. "Lots of things can be bought with a GPC card, like commodities, so for those things there's no need to buy them in-theater" (Shaw AFB Focus Group 2). Items available through the GSA or FSS and large IDIQ contracts were also mentioned frequently as purchase categories that could be managed using reachback.

In the "Contract Size" column, bold type indicates that large contracts above the simplified acquisition threshold of $1 million are also candidates for reachback. For some in Afghanistan, this was a reflection of opinions about the maturity of the local contractor environment and the ability to satisfy large requirements through local vendors.

CCOs generally felt that bureaucratic contracting processes—contract review, legal review, payment processing, and closeout—could be accomplished using reachback because these functions can be performed at a desk, and it does not matter where the desk is. "We were working with people as far away as San Francisco and Phoenix [for legal review]," according to one focus group participant (Wright-Patterson AFB Focus Group 4). However, as the lack of bold typeface or underlining in the "Process" column of Table 3.1 indicates, there was occasional disagreement in the case of closeout: Some people felt that the need to send the contract files elsewhere complicated the use of reachback for this purpose, and others felt that if payments were made locally, closeout was better accomplished locally (Shaw AFB Focus Group 1). Opinions about using reachback for formal or complex source selections were also mixed.[12] According to one focus group participant, "Formal source selections are rare in Afghanistan. For large contracts requiring source selection we'd rather they be done in CONUS" (Wright-Patterson AFB Focus Group 3). On the other hand, we also heard that "big, complex service contracts need someone there in-theater working with the requiring activity to define the requirement."[13] Clearly, the specifics of a given contract and the skill set of the available personnel would affect the decision to use reachback.

Long-term contracts that benefit from the continuity of personnel working with the contract (a stated purpose of the JCC Reachback Branch) and theater-wide purchases that benefit from economies of scale were among other issues mentioned in focus groups and interviews, as shown in the "Other Considerations" column of Table 3.3. We note here three noncontracting issues that were raised when the possibility of reachback was mentioned. First, communication infrastructure (email, telephone, file sharing) must be sufficient to allow fast and reliable communication between theater personnel and the reachback unit. By way of example, the unreliability of communication in some locations in Afghanistan was mentioned as a potential limitation of using reachback. Second, the work hours of the reachback cell must be adjusted to allow responsiveness to everyone involved with a contract. The cell might need to be open late at night U.S. time to communicate with customers many time zones away, but it must also keep regular U.S. hours to respond to contractors based in the United States. Finally, a reachback cell needs to be accountable to the in-theater commander. A common concern of CCOs was that, without some sort of formal agreement, personnel in a reachback organization could be drawn away from in-theater demands by the needs of their local CONUS boss. Even though the JCC Reachback Branch at RICC is not under formal control of JCC-I/A, for example, the memorandum of understanding with JCC-I/A provides for the presence of in-theater liaison officers to improve responsiveness.

There are, of course, reasons *not* to use reachback for contracting. Table 3.4 displays CCO insights into these reasons. The categories listed and the means by which strength of evidence is conveyed are similar to those used in Table 3.3.

---

[12] "Formal source selection is used for high-dollar value or complex acquisitions where someone other than the procuring contracting officer is the source selection authority (SSA). The process begins with the establishment of an evaluation plan for a proposed acquisition, and ends when the SSA selects a contractor to receive a contract award. Informal source selection procedures are less complex, as the contracting officer can determine which offer constitutes best value for the Government without formal input from other government officials specifically designated for that purpose." See Defense Acquisition University, undated (b).

[13] This comment was made in an interview conducted at the Air Education and Training Command (AETC) contracting office at Randolph AFB, February 12, 2009.

Table 3.4
CCO Insight into What Is Not Amenable to Reachback

Requirement	Contract Instrument	Contract Size	Process	Other Considerations
Day-to-day commodities	BPA	Less than $25K	Source selection involving local vendors	Urgent, reactive requirements
Temporary construction			Quality assurance	During initial deployment and buildup phases
Small or local projects			COR duties	
Cement and rebar			Administration of task orders	Maturity of local economy
Anything requiring a site visit				"Iraqi first"/"Afghan first" programs
				Contracts that require close, frequent contact with customer

SOURCES: CCO AARs, 2009 RAND focus groups, and interviews.

NOTES: Standard font indicates thin or mixed evidence for the attribute not being amenable to reachback. **Bold font** denotes attributes that were mentioned frequently or in more than one forum.

Table 3.4 shows that commodities (such as cement and rebar) that are difficult or expensive to transport, small purchases (under $25,000), and items that are needed on short notice (urgent or reactive requirements) are best purchased locally. Processes that require face-to-face contact with vendors or direct observation of contract performance also require someone locally—though not always a CCO.

The presence of BPAs in the "Contract Instrument" column as something not amenable to reachback requires some explanation. BPAs are not technically contracts (Long and Clements, 2007, p. 153). They establish charge accounts with qualified vendors for the purpose of filling anticipated repetitive needs for supplies or services (Long and Clements, 2007, p. 152). While personnel who are not CCOs (ordering officers, for example) can make purchases through BPAs, focus group participants agreed that "someone needs to be there to be responsible for the BPAs. . . . A CCO needs to be in-theater to inspect what is being bought. . . . If you don't have a person there, BPAs can run amok. BPAs need to be managed in-theater" (Hill AFB Focus Group 1).

One other interesting limitation on the use of reachback is the impact of policies related to a particular contingency. The U.S. government has established "Iraqi first" and "Afghan first" policies that have the goal of using local vendors in Iraq and Afghanistan, respectively, to build up the local economy and improve the standard of living. Implementing these policies can limit the flexibility to use reachback for a given requirement or contract.

CHAPTER FOUR

The Impact of Reachback

We have examined the use of reachback by some organizations that provide contracting support for OIF and OEF and have considered the insights of CCOs with deployment experience about what is, and what is not, amenable to reachback for contingency contracting. In this chapter, we apply these insights to data on purchases made by JCC-I/A to explore the potential impact of using reachback on the number of CCOs who were needed to be deployed to support requirements in 2008.

JCCS Contracting Data and Applications

According to JCCS data, JCC-I/A obligated about $7.5 billion in FY 2008 through about 41,400 contract actions. JCC-I/A has about 255 CCOs (Westermeyer, 2009a), so this means that, on average, contracting personnel managed 162 transactions and $29 million per person that year. We use this as a rough metric for how deployments could be reduced if some contracting work were accomplished using reachback. That is, if 162 transactions or $29 million worth of work were moved out of theater, we assume that one less person would be needed in-theater.

Table 4.1 illustrates this potential. The table places FY 2008 spending in several of the categories that are amenable to reachback according to CCOs we interviewed: contracts over $1 million, GPC purchases, GSA/FSS purchases, and commodities. Spending in three other categories—commodities purchased from a nonlocal provider, services provided by a nonlocal vendor, and construction provided by a nonlocal vendor—are also listed. For example, there were 6,406 contract actions over $1 million. Using the rough metric outlined above, about 40 contracting personnel would have accomplished them, so if all of the work related to these actions were moved to a reachback location, perhaps 40 fewer personnel would need to be deployed to JCC-I/A.

Based on the dollar value of these contracts, $5.9 billion, 205 people would have worked on them, so moving the contracts to a reachback location could have a large effect on the number of deployed personnel.[1]

---

[1] Although this may seem unreasonably large, note that 79 percent of the contract dollars in FY 2008 were for contracts larger than $1 million.

Table 4.1
JCCS Data on Spending in Various Categories, FY 2008

Type	Contracts over $1M	GPC	GSA/FSS	Commodity	Commodity (nonlocal provider)	Service (nonlocal provider)	Construction (nonlocal provider)
Contracts (% of total)	649 (2%)	1,666 (6%)	1,594 (6%)	19,195 (73%)	9,832 (37%)	987 (4%)	337 (1%)
Actions (% of total)	6,406 (15%)	1,754 (4%)	1,912 (5%)	26,244 (63%)	12,028 (29%)	3292 (8%)	706 (2%)
Dollars (% of total)	$5.944B (79%)	$4.3M (.06%)	$385M (5%)	$1.857B (25%)	$707M (9%)	$3.105B (41%)	$346M (5%)
Potential reduction in deployed personnel	40 to 205 (40 based on actions, 205 based on dollars)	0 to 11 (0 based on dollars, 11 based on actions)	12 to 13 (12 based on actions, 13 based on dollars)	64 to 162 (64 based on dollars, 162 based on actions)	24 to 74 (24 based on dollars, 74 based on actions)	20 to 107 (20 based on actions, 107 based on dollars)	4 to 12 (4 based on actions, 12 based on dollars)

SOURCE: FY 2008 JCCS data, downloaded from the JCCS website on January 22, 2009.
NOTES: Contract value thresholds were determined within a given fiscal year. The difference in total dollars between this method and one not constrained by fiscal year to identify contracts meeting the threshold (i.e., considering contract value overmultiple years) is about 1 percent.

If we use the lower values for potential deployment reductions shown in the last row of Table 4.1 for the first four categories, the total potential reduction is 116 (40 + 0 + 12 + 64). Although this makes the unlikely assumption that none of these contract categories overlap and workload is evenly distributed across CCOs, it does illustrate how estimates can be made.[2]

We use this approach to address the potential impact of deployment reductions on the deploy-to-dwell time of the Air Force contracting career field (64P for officers and 6CXX for enlisted personnel). In FY 2008, with about 286 deployed positions worldwide, this career field had a deploy-to-dwell ratio of 1:1, which implies that only 572 personnel were available to deploy.[3] Table 4.2 describes the situation.

The "Current" column of Table 4.2 shows the current state that leads to a 1:1 dwell ratio. To reduce the deploy-to-dwell time to 1:2 (six months deployed and one year home, for example), either the number of deployed positions needs to be reduced or the number of available personnel needs to be increased. With 572 personnel, a 1:2 dwell time would imply 191 deployed positions (572 divided by 3)—95 fewer than currently exist. As we just saw, such a reduction might be possible if all work in four categories (contracts larger than $1 million, GPC, GSA/FSS, and commodities) were moved to a reachback location (again, assuming no overlap in the contracts involved and an even distribution of workload).

---

[2] Overlap clearly exists: Adding up the higher values for potential reductions in the fourth row gives a number higher than the number of people actually deployed. A more detailed analysis of overlapping contracts or a workload analysis such as those accomplished in the development of Air Force manpower documents (see USAF, 2001) would be required to accurately determine reductions.

[3] With 572 personnel available to deploy and a 1:1 deploy-to-dwell ratio, 286 would be at home and 286 would be deployed at any given time. We use this implied figure because SAF/AQC was unable to provide the number of personnel actually considered available to deploy.

Table 4.2
Deployment Reductions Required to Affect
Dwell Time

Availability	Current	Potential
Deploy-to-dwell time	1:1	1:2
CCOs available to deploy	572[a]	572
Deployed positions	286	191

SOURCES: Deployed positions based on Cameron, 2008, and Westermeyer, 2009a.

[a] The number 572 available to deploy is implied by the FY 2008 number of deployed positions (286) and the current 1:1 deploy-to-dwell ratio.

We emphasize that this analysis is just a starting point. CCO focus groups made it clear to us that many factors affect the potential to use reachback for an individual contract. Additionally, the Air Force could not make this reduction on its own because JCC-I/A positions are joint billets, and their elimination would have to be negotiated through the process that results in the joint manning document.[4]

Variation in Reachback Potential with Time and Location

CCO comments in focus groups and interviews made it clear that opportunities to use reachback for contracting functions vary with the phase of the contingency and with the location of the contingency. Table 4.3 illustrates this variation by comparing data from Iraq in FY 2003, when the contingency began, to data from FY 2008.

The circled cells in the table show the huge increase in the number of actions related to contracts that were above $1 million as the situation in OIF evolved to a sustainment phase— from only 50 actions in 2003 to 4,389 in 2008.[5] Since CCOs consider contract actions associated with contracts over $1 million amenable to reachback, this increase represents the potential to reduce the number of deployed personnel.

Table 4.4 (with the same entries as Table 4.3) makes an analogous comparison based on location. We see that even though operations in Afghanistan have been conducted longer than those in Iraq, spending is "less mature" in the sense that fewer contract actions are related to large contracts. As shown in the column labeled "2008 Afghan Contracts Above $25K," only 19 percent of contract actions in Afghanistan are associated with contracts larger than $25,000, while 36 percent of the actions in Iraq are (as shown in the column labeled "2008 Iraq Contracts Above $25K"). These numbers imply that more time is spent on smaller, day-to-

---

[4]   The process of developing a joint manning document is described in Chairman of the Joint Chiefs of Staff Instruction (CJCSI) 1301.01C, 2004. The document does say that, when validating these positions, the Joint Staff should "Ensure maximum use of reach back, contractors, and centralized joint organizations (e.g., regional contracting and intelligence centers)" (CJCSI 1301.01C, Enclosure A, para. 4.c[3]).

[5]   The standard phases of a contingency are initial deployment, buildup, sustainment, and termination/redeployment. Note that the percentage of contracts above $1 million decreased from 2003 to 2008. This could be the result of the Iraqi First program, which increased the number of low-dollar contracts to local vendors.

day purchases, which are among the purchases we saw in Table 3.2 that may not be amenable to reachback.[6]

Table 4.3
Spending Variation in Iraq, 2003–2008

Type	2003 Iraq Contracts Above $25K	2003 Iraq Contracts Above $1M	2008 Iraq Contracts Above $25K	2008 Iraq Contracts Above $1M	2008 Afghan Contracts Above $25K	2008 Afghan Contracts Above $1M
Contracts (% of total)	93 (58%)	26 (16%)	7,378 (27%)	507 (2%)	3,116 (12%)	144 (0.5%)
Actions (% of total)	134 (65%)	50 (24%)	14,757 (36%)	4,389 (11%)	7,995 (19%)	2,017 (5%)
Dollars (% of total)	$447M (99.9%)	$436M (97%)	$5.911B (79%)	$4.885B (65%)	$1.493B (20%)	$1.059B (14%)
Potential reduction in deployed personnel	1 to 15 (1 based on actions, 15 based on dollars)	0 to 15 (0 based on actions, 15 based on dollars)	89 to 204 (89 based on actions, 204 based on dollars)	27 to 168 (27 based on actions, 168 based on dollars)	49 to 51 (49 based on actions, 51 based on dollars)	12 to 37 (12 based on actions, 37 based on dollars)

SOURCE: FY 2008 JCCS data, downloaded from the JCCS website on January 22, 2009.

Table 4.4
Spending Differences Between Iraq and Afghanistan

Type	2003 Iraq Contracts Above $25K	2003 Iraq Contracts Above $1M	2008 Iraq Contracts Above $25K	2008 Iraq Contracts Above $1M	2008 Afghan Contracts Above $25K	2008 Afghan Contracts Above $1M
Contracts (% of total)	93 (58%)	26 (16%)	7,378 (27%)	507 (2%)	3,116 (12%)	144 (0.5%)
Actions (% of total)	134 (65%)	50 (24%)	14,757 (36%)	4,389 (11%)	7,995 (19%)	2,017 (5%)
Dollars (% of total)	$447M (99.9%)	$436M (97%)	$5.911B (79%)	$4.885B (65%)	$1.493B (20%)	$1.059B (14%)
Potential reduction in deployed personnel	1 to 15 (1 based on actions, 15 based on dollars)	0 to 15 (0 based on actions, 15 based on dollars)	89 to 204 (89 based on actions, 204 based on dollars)	27 to 168 (27 based on actions, 168 based on dollars)	49 to 51 (49 based on actions, 51 based on dollars)	12 to 37 (12 based on actions, 37 based on dollars)

SOURCE: FY 2008 JCCS data, downloaded from the JCCS website on January 22, 2009.

---

[6] Differences between contracting in Iraq and Afghanistan could be the result of many factors, such as the commander's concept of operations, the number of main operating bases, the number of forward operating bases, the number and types of personnel, and the availability of local providers.

Advantages of Reachback Beyond Deployment Reductions

Tables 4.1 through 4.4 have shown how the use of reachback could potentially reduce the number of contracting personnel needed to deploy. Notwithstanding deployment reductions, however, reachback might be useful for other reasons.[7] Among the advantages most commonly mentioned in our focus groups were continuity of the workforce managing the contract, standardization of contracts that satisfy similar requirements, reachback organizations as "repositories" of contract expertise, and reachback as a training opportunity for inexperienced contracting personnel.

The regular turnover of contracting personnel who are on six-month deployments means that arriving contracting officers have a limited amount of time to learn about existing in-theater contracts from the people they are replacing. Managing long-term contracts through a reachback organization with a stable contracting workforce could mean that new contracting officers would not require periodic retraining on the details of an existing contract.

> If you had a reachback cell for larger items, more people would be familiar with all the contracts already there. You could piggyback on an existing contract rather than create your own. For example, MRAP [mine-resistant armor protected] vehicles. There was already a contract in place in one area to get them, but no one knew that in-theater. If that were reached back, it would be known to everyone. You don't know about and don't have time to find out about supply contracts in place. (Randolph AFB Focus Group 2)

> With military contracting officers you lose the continuity. Most of our customers at AFCENT are civilians, so it helps to work with civilians [in the reachback location] and build long-term relationships. You don't have to worry about civilians deploying. You can base the workload off of stable manning. Right now, I have 11 here today, and then it goes down to 9. It's [that is, a civilian office] a more stable office. With operational contracting after the next rotation, you could have enough now and then not. That increases the burden on others. (Shaw AFB Focus Group 3)

Contracting officers in a reachback organization that deals regularly with similar requirements from the theater may also be in a better position to pool requirements from different customers to take advantage of economies of scale or to leverage business with large vendors.

Some officers we interviewed noted that formal source selections were accomplished only rarely in some places and that deployed personnel did not always have the experience or skills to conduct one. In such cases, a reachback organization with long-term, highly skilled personnel could be a resource to provide expertise for complicated contracts when necessary.[8]

Finally, focus groups indicated that reachback organizations provide opportunities for improved CCO training. By learning about how to deal with contracting issues in the AOR while in a low-threat reachback location, inexperienced contracting officers could be better

---

[7] In fact, the use of 20 CONS resources for reachback did not mean that fewer CCOs were deployed, nor did it mean that the contracting squadron gained personnel to do the extra work. LOGCAP is viewed by some Air Force CCOs as an added drain on resources because Air Force CCOs are assigned to DCMA to serve as administrative contracting officers for LOGCAP (Wright-Patterson AFB Focus Group 3).

[8] Ross, 2009, presents results of a competency assessment that was conducted from January to July 2008. Slide 13 of the briefing notes, "It has long been the perception that the AF contracting workforce needs extensive formal source selection training."

prepared for when they actually deploy. While not completely disagreeing with this comment, other contracting officers we interviewed felt that training was better accomplished the other way around: Inexperienced personnel deployed in-theater would gain experience that would make them better able to function effectively in a reachback cell because they would fully understand the importance of what they were doing and the difficulties experienced by those in-theater. Table 4.5 displays representative focus group comments related to these advantages of reachback.

Potential Disadvantages of Reachback

Support for using reachback for some contracting functions is not unqualified. Despite the potential for reductions in deployments and the possibility that reachback might improve the efficiency and effectiveness of some contracting functions, there are potential disadvantages. First, some CCOs are skeptical that using reachback will actually reduce deployment demands for Air Force CCOs. Once a "body" is in-theater, they explained, a commander is unwilling to give it up, because there is almost always something that needs to be done—even if it is not directly related to contracting.

Contracting officers also pointed out that using reachback may not reduce a CCO's workload. We noted earlier in this report that tasks deemed low priority, such as contract closeout, are often not accomplished; hence, using reachback to perform low-priority tasks will not mean that CCOs in-theater will have less to do. Further, since in-theater CCOs are recognized as the contracting experts, customers and commanders depend on them for a variety of contract-related tasks, even for contracts for which they are not technically responsible. For example, we heard in focus groups that in-theater CCOs were asked to help with LOGCAP- and AFCAP-

Table 4.5
Potential Advantages of Using Reachback

Continuity	Standardization
I had a reachback counterpart. . . . It helped with continuity on services contracts. When we want to contract something, we can go back and look at the files and see who got the previous contract and how it was set up. That can be done stateside. (Hill AFB Focus Group 2)	There's a benefit of reachback that by pooling contracts you can get standardization, which would also help projects that have specific requirements. (Wright-Patterson AFB Focus Group 2)

Reachback as repository of expertise	Training
Formal source selections are rare in Afghanistan. For large contracts requiring source selection, we'd rather they be done in CONUS. (Wright-Patterson AFB Focus Group 3) You spend lots of time training [contracting officers]. You can get some "knowledge pool accumulation" [if you keep highly trained officers stateside, working together] and will have a better product. (Hill AFB Focus Group 2) You would get a mix of skills with reachback [that you don't necessarily have in the field]. (Hill AFB Focus Group 2)	One side effect might be improved CCO training. [By working stateside in a reachback cell,] they can get to know how the AOR works a little even if they aren't over there. You could improve overall training, they could learn procedures through reachback before being deployed in-theater. Also, you can have access to proven practices, to expertise you can't get over there. It could be a training bed. (Randolph AFB Focus Group 2)

SOURCE: 2009 RAND focus groups.

related issues even though they did not manage the contracts. A fear is that if reachback is used for other contracts, the local CCO will still be expected to provide assistance with them, and so his or her workload will not be reduced.

Related to this comment is the possibility that if reachback is used and the number of deployed CCOs is reduced, those CCOs still deployed will experience greater stress because the workload will not actually have been reduced commensurate with the reduction in personnel. Civilian contracting personnel pointed out that when 20 CONS was performing reachback for USAFCENT, their stress level also increased because of the larger workload without a larger staff.

Finally, there is the often-stated concern by Air Force personnel that other services need to increase the size and improve the training of their own contracting career fields before the Air Force will be relieved of the contracting burdens it now bears: As long as Air Force contracting personnel fill 70 percent of JCC-I/A billets (Westermeyer, 2009a, slide 11), reachback may not provide that much relief.[9] Table 4.6 shows typical comments from focus groups related to these issues.

Summary

This chapter has demonstrated how to use JCCS and focus group data to explore the potential impact of using reachback to reduce the number of deployed personnel. One of the surprises from our focus group discussions, however, was that reachback was not seen as the

Table 4.6
Potential Downsides of Using Reachback

It may not reduce total AF CCO demands	It may not reduce workload
We've so often seen this. . . . The second you say "yes" and commit to the number of officers, that number is never allowed to go down. . . . Reachback has definite potential to reduce day-to-day strain, but there will just be more things on your plate. (Wright-Patterson AFB Focus Group 3) If there's a greater capability, something will fill it in. We do the job well; consequently there's never an end to the work we're asked to do. (Shaw AFB Focus Group 1)	CCOs are the focal point for any contracting issue. . . . So everyone comes to us. We get calls about deliveries and have to send people over to DCMA. (Hill AFB Focus Group 2) We had issues with LOGCAP, and people would come to us because we're [the] contracting officer. . . . It sometimes got really complicated. Customers came to us first. (Shaw AFB Focus Group 1)
It could increase workload stress	**Other services must do their part**
If you reduce the workload and then reduce the number of people deployed, that doesn't necessarily reduce stress on the people still in-theater. Reducing the number of people deployed [by] too much will kill those still in-theater. (Randolph AFB Focus Group 2)	Get the Army to step up [and provide some contracting officers of their own]. You won't fix the AF CCO tempo until the Army fixes theirs. (Wright-Patterson AFB Focus Group 2)

SOURCE: 2009 RAND focus groups.

---

[9] The first recommendation of the Gansler Report was that the Army "increase the stature, quantity, and career development of military and civilian contracting personnel (especially for expeditionary operations)" (Gansler, 2007, p. 47). In February 2009, each military department and DoD agency was directed to conduct a total force assessment of their required CCOs and CORs (Assad, 2009). Actions the Army has taken in response to the Gansler Report are described in R. Thompson, 2009.

most important solution to effectiveness or stress problems related to deployment. In the next chapter, we examine other factors that cause stress in the contingency environment—some of which are not affected by the use of reachback.

CHAPTER FIVE

Other Issues Related to CCO Stress and Contracting Efficiency

The CCOs with whom we spoke expressed some interest in the potential of reachback to decrease in-theater workload and stress, but they also pointed out that there were other factors that contributed to stress that were not necessarily related to, or alleviated by, the use of reachback. Although some focus group participants were able to devote most of their time to buying, others spent considerable time on different tasks that they believed could or should have been accomplished by someone else. This chapter discusses task- and workload-related problems encountered by deployed CCOs that add to their stress and decrease contracting efficiency but that require approaches other than reachback as a solution.

Task-Oriented Problems

CCOs' perceptions of their deployment experience are valuable as an indicator of their morale or job satisfaction. By shedding light on any deployment frustrations that CCOs experience, their accounts provide Air Force leadership with an opportunity to address sources of such discontent via reachback or other means, with potential favorable implications not only for CCOs' morale but also for their retention. Accordingly, we coded the AARs not only for references to contracting tasks, as discussed in Chapter Two, but also for references to contracting-related *problems*. This enabled us to consider how often CCOs described tasks in the context of problems and to understand what those problems were. Figure 5.1 depicts the results of our analysis.[1] The light gray bars show the number of AARs that were coded for a reference to a specific type of task, and the dark gray bars indicate the subset of those same AARs in which contracting tasks were framed by a CCO as problematic.[2] For example, we identified 88 AARs

---

[1] We did not perform a statistical analysis of the differences in problems reported by location because we were unsure how representative the AARs were of *all* CCOs who may have been deployed. A look at the raw data hints that some differences may exist: There was a greater proportion of problems related to presolicitation tasks in Iraq than in Afghanistan (29 percent of Iraq AARs that reported problems were in this category versus 15 percent of those for Afghanistan), and there was a lower proportion of delivery coordination–related problems in Afghanistan (12 percent of those that reported problems) than in Iraq (20 percent of those that reported problems).

[2] AFFARS Appendix CC (USAF, 2007), instructs CCOs to include in their AAR "problems encountered with the contracting process," and this may contribute to the high percentage of AARs that mention problems. However, not only did we observe differences in the frequency with which different problems were cited, but also, when problems are mentioned, they tend to be very specific. Here is one example:

> Insurgents were killing vendors working with the coalition forces or kidnapping them and demanding ransom. . . . Many times the CCO had to coordinate convoys to alternate delivery locations to pick up goods from vendors. The process was lengthy and took many man-hours to complete which took away from the Contracting mission. (Iraq AAR)

Figure 5.1
Frequency with Which CCOs Discussed Tasks in the Context of Contracting Problems

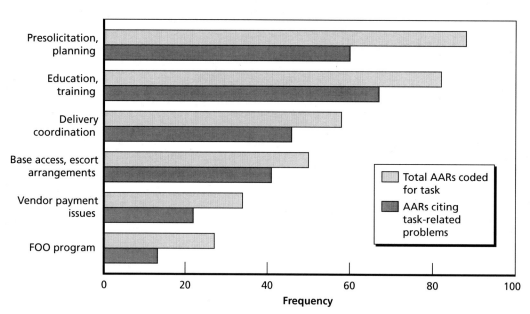

SOURCE: CCO AARs.
NOTE: N = 236 AARs.
RAND *TR862-5.1*

as having a discussion about presolicitation or planning tasks, and in about two-thirds (60) of those AARs, presolicitation tasks were part of a problem-oriented discussion.[3] The following comments are taken from those AARs:

> While many Air Force personnel are aware of Contracting's mission and purpose and may even have worked with Contracting in some capacity (QAP [quality assurance person], BPA caller, GPC cardholder, assisted in writing a SOW), the typical Army customer in a deployed environment has little or no experience working with Contracting. This creates additional challenges for the CCO in terms of defining requirements, writing a SOW, and interfacing with contractors. (Afghanistan AAR)

> [T]he lack of quality products from CE (Civil Engineering) was the biggest problem throughout the rotation. Engineers expected to be able to fix SOW problems during the site visit or correct design flaws during pre-construction conferences. Because the CCOs did not want to be seen as the hold-up, they frequently spent countless hours fixing products CE should never have submitted. (Iraq AAR)

> It would have also helped to have a dedicated engineer embedded with the CO to help define and write Statements of Work (SOW). On more than one occasion I received requirements from the field without a SOW attached. In turn, I had to write the SOW to the best of my ability. (Iraq AAR)

---

[3] Many of the problems were associated with requirements or statements of work: Twenty-six of the 60 AARs (43 percent) that refer to problems related to planning and presolicitation tasks discussed problems specifically about requirements and/ or SOWs.

A review of Figure 5.1 also reveals that education, a type of contracting operations oversight, and two aspects of post-award contract administration—delivery coordination and base access—were mentioned in the context of problems roughly four-fifths of the time they were described. Again, we provide excerpts from the AARs, in this case to illustrate how CCOs discussed their education efforts in relation to contracting problems:

> I needed to train and educate my customers as much and as often as possible with respect to their roles and responsibilities, to cut down on mistakes and to help them understand the importance of a well-defined requirements package. (Iraq AAR)

> Continuous acquisition process education was necessary for the majority of our customers. Prior to deployment, many customers had never generated requirements and/or performed QAE/inspector duties. (Kyrgyzstan AAR)

Remarks about delivery coordination and base access tasks that were shared while outlining problems include the following:

> As long as a continuous terrorist threat remains in the area, getting contractors and deliveries to Ali in a cost effective and timely manner will be an issue. With potential violence in the local area between coalition forces and insurgents, IEDs [improvised explosive devices], and the random kidnapping of contractors, the logistical challenges remain. While the industrial park alleviates a portion of this problem for base service contracts, ECONS must continue to work with the Army on convoy issues and better long-term planning with customers to accommodate 2–4 week delivery times on most items (APO, TMO [Traffic Management Office]). (Iraq AAR)

> Dealing with Qatar Customs was an administrative burden on the contracting officer. The whole clearance process was laden with bureaucracy, red tape, and had no clear-cut procedure because it was subject to change without prior notice. The absence of a SOFA [Status of Forces Agreement] or Host Nation Agreement compounded the situation. (Qatar AAR)

> All personnel E-6 and below were required to pull a day of escort duty several times during their rotation. Officers and E-7s were required to be escort supervisors. Inclusion of CCOs as part of escort details resulted in numerous logistical difficulties and contract administration problems. (Kyrgyzstan AAR)

> When the contractors were awarded a contract and getting their equipment on base, it could take as long as four hours to get everything on base. During site visits, the construction CO developed a work around that seemed to speed up the process. He would have contractors leave their vehicles outside the base and simply process themselves through the base Pass and ID section and then he arranged transportation. The Construction CCO would herd them, as many as 18, contractors onto a van or bus with armed escorts trailing behind. (Iraq AAR)

Although cited less frequently as a problem than other post-award contract administration tasks, vendor payments were still clearly a concern for CCOs: About two-thirds of the time that payment-related tasks were described in the AARs, it was in relation to a challenge of some sort. CCOs tended to discuss vendor payment issues, such as the following:

The banking system in Afghanistan is not very mature, and occasionally local vendors would complain about not being paid, even though the Army Finance office had copies of relevant EFT [electronic funds transfer] transmittals. Often, the banks would resolve the payment issues after being presented with the documentation of the EFT transmission; however, these problems cause an additional work load on both contracting and finance. (Afghanistan AAR)

All payments are initially sent to VRCC [Camp Victory Regional Contracting Center] for processing. Once payment packages were assembled per the instructions above, it was sent to Finance for them to process. Finance officially has 48 hours to review a package and return it to the CCO if budgeting had not obligated the funds or more information was missing/incorrect. This process has become a bit easier; however it initially took up to two weeks for finance to review a payment package. It was then up to the CCO to contact budgeting, the customer, or the vendor in order to correct any issues. (Iraq AAR)

DFAS [Defense Finance and Accounting Service] Payment Issues: There was a tremendous backlog of payments at the beginning of the rotation. DFAS-Rome was the payment office for the Afghanistan AOR. Many times it was difficult or impossible to find the equipment ordered. Many times the customer would have rotated and the replacement had no knowledge of the order. Some payments were more than a year overdue. Each site established a weekly telecom with DFAS to work through the backlog. (Uzbekistan AAR)

Finally, roughly half of the discussions about administering the FOO program also contained references to FOO program challenges, such as the following:

FOOs that were located at Forward Operating Bases (FOB), many times were only able to clear when they ran low on available funding and were required to return to Kandahar Airfield to draw more funds. The problem that this created was that it was often hard to correct discrepancies or mistakes with purchases at an early date so that they could be corrected for subsequent purchases. (Afghanistan AAR)

Clearing Field Ordering Officers (FOO): VRCC cleared FOOs once a week on Tuesdays on a walk in basis. Although the act of clearing a FOO was not difficult there were many problems with FOO personnel not following guidelines and regulations on their transactions. FOOs were turned away for corrections or violations were corrected at the time of clearing if possible. (Iraq AAR)

Time-Consuming Tasks

One type of problem that spanned task categories was the perception held by some CCOs that certain tasks were more time-consuming in deployment than at home; as a result, accomplishing them meant that other tasks were overlooked. Although data were not available to describe precisely how much time CCOs spent on different tasks while deployed, both the AARs and focus groups provide insights about what CCOs perceived to be time-consuming—perhaps surprisingly so, based on their understanding of CCO duties. Specifically, when asked about time-consuming tasks, focus group participants emphasized the following as requiring a considerable amount of time—most of which have already been noted as problem areas for other reasons:

- customer requirements and SOWs[4]
- education and training
- base access, escort arrangements
- delivery coordination
- vendor payments
- the FOO program.

Supporting evidence from the focus groups, along with corroborating examples coded in the AARs, is provided in Table 5.1. Further, as this comment from a CCO deployed to Jordan suggests, at times, CCOs had multiple tasks that were not only time-consuming but also potentially viewed as outside their purview:

> In addition to contracting responsibilities, my "other" day-to-day responsibilities consumed an equal or greater amount of my time and sometimes created potential conflict of interests. "Other" responsibilities included (1) acting as the J4 POC [point of contact] in country planning, determining, validating and coordinating all new requirements; (2) acting as my own Paying Agent; (3) receiving and delivering all goods to the mission area and tracking disposition of every item; and (4) performing budget and financial actions, to include creating and managing a budget, reconciling and distributing contract and other financial documentation, and coordinating and resolving budget issues between the Comptroller (Qatar) and the MAP [Military Assistance Program] Financial Specialist. (Jordan AAR)

Particularly during focus groups, CCOs suggested that time spent on the aforementioned tasks, as well as on other "higher-priority" tasks, such as contract awards and procurement, detracted from other CCO responsibilities—namely, tracking-related tasks and contract closeout. As the following remarks demonstrate, focus group participants indicated that the required documentation efforts were at times neglected:

> Documentation wasn't getting done. We would look in older files for what we needed to know and wrote it down on scrap paper. If we got stuck and needed to know who the customer was and contact, we couldn't ask anyone because the previous person had redeployed back to the states. We had no documentation on the last person the CCO talked to. Technically mandatory stuff was sometimes neglected. . . . The wing commander doesn't care if contracts are documented. No self-inspection occurs. No one notices what we can't get done. (Shaw AFB Focus Group 1)

> Online actions for GPC holders fall by the wayside. For each GPC buy, we had to update four databases! (Wright-Patterson AFB Focus Group 1)

---

[4] Poor descriptions of requirements and inadequate SOWs from customers are typical complaints by contracting officers in CONUS, but in CONUS there is usually more time for the contracting officer to explain what is wrong and have the customer make changes. (This comment was made during an interview at SAF/AQC on August 17, 2009).

Table 5.1
Examples of Remarks About Time-Consuming Tasks

Task	Remarks
Customer requirements and SOWs	[A]nother thing we spent a lot of time on was acquisitions requirements development, walking through the process [with the customer], who had no ability or desire to define requirements. I would say I spent about 25 percent of my time on requirements development. . . . (Wright-Patterson AFB Focus Group 2)
	Army requirements are often not well-drafted, are not ready to go, and it takes two weeks to define requirements that are currently undefined. This is different than the Air Force. With the Army, many times they [requirements] are not properly defined. We spend a lot of time working with the customer to define requirements. (Wright-Patterson AFB Focus Group 4)
	SOWs coming from customers were not adequate. A lot of time was spent sending SOWs back and forth for corrections. Customers tend to put a lot of the contractual terms and conditions in the SOW, even after being told this was not proper. If not caught by the CCO, this creates conflicting language in the awarded contract. (Kyrgyzstan AAR)
Education and training	The customer cannot put together a package, so you go with them and show them how, and it eats up a lot of your time. (Randolph AFB Focus Group 1)
	You spend a lot of time training CORs, commanders, CACI employees, GRD [Gulf Regional Division]. . . . They don't come to the theater trained [for contracting]. (Hill AFB Focus Group 2)
	A lot of my time and aggravation was spent educating personnel on proper contracting procedures. (Thailand AAR)
	Customer Education was continuously required. This was due to the frequent turnover rates of government personnel and lack of education to the procurement process. Educating customers proved to be very time consuming throughout the deployment, but once the customer was educated on the process it proved to pay off with clear requirements and ease of the contract thru delivery. (Iraq AAR)
Base access, escort arrangements	We had to go and argue with the gate guy to get the guy on base. You show him your badge, you had to sign some paper. It takes a lot of time. (Randolph AFB Focus Group 1)
	In Al Dhafra, we spent 70 to 75 percent of the time trying to get contractors onto the base. (Hill AFB Focus Group 1)
	Getting contractors through the Army checkpoint and through the Air Force Search Pit proved to be our biggest headache and demanded significant time and coordination. (Iraq AAR)
	FYI [c]ontracting seems to be the POC for all security related issues; security clearance procedures can be and are very time consuming. (Pakistan AAR)
Delivery coordination	We spent a lot of time following up and finding out where things were being delivered. Things arrive and you have to get it out; you had to make daily calls to coordinate delivery. (Wright-Patterson AFB Focus Group 3)
	I would then follow up deliveries, customs, visits to the ministry of interior, etc. We followed up with payments from DFAS. Customers only care if they don't have a Standard Form 1449 for inspections of delivery. At Al Udeid we did clearances and policy review. I was a shoulder to cry on in the field. Customs was taking lots of time. (Shaw AFB Focus Group 1)
	Contracting operates under a cradle-to-grave concept. Although this is fine, we usually spend a lot of time coordinating for vendor deliveries with TCN [third-country national] escorts. (Iraq AAR)
	Local customs was approximately 65 percent of my duties. While not a huge problem, it was a daunting task trying to get the extremely wide variety of items through the required customs clearance process. The massive amount of paperwork attributed to the customs procedures was just one of the repetitiously tedious tasks. (Afghanistan AAR)

Table 5.1—Continued

Task	Remarks
Vendor payments	Vendor payments take time. EFT payments are tough. Contractors in Iraq aren't sophisticated enough to manage it. They didn't understand it. Most of them haven't even ever seen a credit card in their life, much less know what an account routing number is. (Wright-Patterson AFB Focus Group 3)
	I know what took time—payments, escorting contractors onto base, and getting them paid. (Randolph AFB Focus Group 1)
	The lion's share of each CO and KO's [contracting officer's] day is taken up with making payments to the local contractors, leaving the evening and night as the only suitable time to solicit, construct, award, and close out. (Iraq AAR)
	With multiple charges to the same vendor on one card, proving that a new charge had not already been paid was very difficult and time consuming. (Iraq AAR)
FOO program	You end up spending time training FOOs. (Hill AFB Focus Group 2)
	[Any] time they came in, we would clear them. But the job became unmanageable with 500 to 600 FOOs across the FOBs. So we would have "FOO day" on Tuesday and we would split the office in half. One half to clear them in the morning, the other half to clear them in the afternoon, so we could all get at least some work done. I'm not saying this shouldn't be done; there does need to be oversight, but it's a big drain on time. (Randolph AFB Focus Group 2)
	Field Ordering Officer training at the Phoenix Academy [special training for U.S. military personnel] is a continuous issue because the individual in charge of scheduling dates did not coordinate with all organizations to establish the training days. We trained an average of 200-plus members on a monthly basis. This training would take an individual away from their other duties for a minimum of at least two hours per briefing. (Iraq AAR)

SOURCES: CCO AARs; 2009 RAND focus groups.

> In my office, the guys had no time to update JCCS, and so since nothing was being updated, they [higher-ups] thought we had no work to do and so they pulled people from our office. (Wright-Patterson AFB Focus Group 1)

The last remark suggests that failing to update contract databases may not only make it difficult to track shipments, verify payments, and perform other post-award functions, but it may also have manpower implications.

As implied earlier, focus group participants also felt that contract closeout was of lower priority than efforts directly related to providing customers with the goods and services they needed (e.g., awards, delivery coordination, performance monitoring). The following excerpts help convey their viewpoint:

> Closeouts are not a priority in-theater, so it's already not getting done. Taking it back stateside doesn't reduce the burden. Making the files look pretty is not a priority. (Hill AFB Focus Group 1)

> Contract closeouts fall way behind until eventually "the hammer falls" and we have to spend time closing out a bunch of old contracts. That was our Christmas present in Iraq: we got to spend Christmas Eve closing out old contracts. Anything that was still open had to be done. Granted, it should be as much a part of the process as anything else, but that's a hard standard to reach when you've got 25–30 [new] contracts in the hopper at any given time. (Wright-Patterson AFB Focus Group 2)

The unofficial status of contract closeouts as a lower-priority task was also salient to some AAR authors:

> Close outs typically fell to the bottom of the priority list. (Iraq AAR)

> Due to the deployed operational tempo and so few contracting officers available, contract follow-up and close out was extremely difficult to accomplish. The sheer volume of requests from so many different agencies caused the backlog of closeouts to increase while the entire focus of the purchasing contracting officers was on completing buys and working problems to get customers the equipment they needed when they needed it. (Iraq AAR)

CCO Workload Challenges

CCOs recognize their role as a business adviser to the local commander, but the work they are doing in addition to buying things for the warfighter—such as requirements specification, base access and escorts, payment processing, and quality assurance—can make them feel overworked, especially when they perceive that there are others with lighter workloads who could be available for some of the "extra" duties placed on CCOs.

> Contracting is easily one of the busiest offices on the base. We need to focus on building statements of work, timelines, and also on QA at the back end. . . . After you get done working a 12–15 hour day and you go for your one hour at the gym to decompress and you see all the brackets for team sports all over the place because no one is working as long as you are, you start to stop caring. (Wright-Patterson AFB Focus Group 3)

We also note that, even though a primary goal of this research was to explore the possibility of decreasing the number of CCOs deployed, some of those we interviewed believed that too few were deployed now and that the distribution of those deployed needed to be changed because of imbalances in workload.

> This was a big complaint in Iraq because they are severely undermanned in Iraq. Manning is a big problem. Balad is undermanned. The original manning levels need to be re-evaluated and looked at across locations to determine if they are still appropriate. Some people [CCOs] don't work hard and others work like dogs. (Randolph AFB Focus Group 1)

In addition to the poor distribution of manpower, some CCOs expressed concern about the poor distribution of skills. For example, officers and senior enlisted personnel were sometimes performing contracting tasks that could easily have been done by lower-ranking airmen—airmen who could benefit from the experience. "Even for administration . . . you have O3s running service contracts, and you could have senior airmen doing that instead" (Wright-Patterson AFB Focus Group 2). As a contingency moves to the sustainment phase, the skill set required by CCOs changes (for example, experience with formal source selections can become more important), and there was also concern about the proper development of CCOs.

> A lot of what I did was major source selections, sometimes millions or billions of dollars. I didn't learn that doing operational contracting. We are getting systems contracting as we speak. The skills that I used and needed were not developed by doing small construction contracts and AFCAP-like contracts. I don't think we're growing enough officers to do that mission, and not just for war. (Wright-Patterson AFB Focus Group 4)

> We can have lieutenants and airmen in-theater to do things like closeout and administration. That would be all they do [i.e., no buying], and they would get exposure [to the contingency environment]. I have lots of experience. I can really dive in and get into it. Bring me someone and I'll educate them on the small stuff, on specific tasks and problems, so that on their next deployment, they know what some of the problems are. It would be a great education, so when they go out to be a buyer, they will already be familiar [with the setting] and won't be overwhelmed. (Randolph AFB Focus Group 2)

Finally, there was the oft-expressed view that, as a whole, the Air Force contracting workforce was doing more than its share in-theater, an opinion corroborated by the high percentage of Air Force CCOs in JCC-I/A.

Summary

In this chapter, we used CCO data from AARs and focus groups to highlight three problem categories that contribute to the stress of the contingency contracting environment. The first relates to phases of contracting, such as presolicitation planning, that, for a variety of reasons, seemed more difficult than in a noncontingency environment. The second relates to tasks that CCOs expect to accomplish, such as developing SOWs and providing education and training, but are more time-consuming in-theater. The third category relates to a heavier workload, both because of unexpected duties (such as serving as escorts for contractors) and because of inappropriate distributions of personnel and skills.

CCOs in our focus groups recognized that early in a contingency they might be asked to perform contract-related duties that would normally be accomplished by someone else—the joint handbook for contingency contracting warns them as much:

> During [the mobilization and initial build-up] phase, CCOs may find themselves in the undesirable position of being the requestor, approving official, certifying officer and transportation office for deliveries since the CCO must be prepared to award contracts immediately upon arrival at the deployment site. (Long and Clements, 2007, p. 101)

However, as locations become better established and operations approach the sustainment phase, CCOs feel that responsibilities should be better distributed so they can "focus on the business advisor role, file documentation, cost reduction, and other efficiencies" (Long and Clements, 2007, p. 104). Otherwise, time constraints can lead to CCOs neglecting low-priority contracting tasks, such as database updates and contract closeout. As one CCO said,

We do a lot of things that are not really our job. We take ownership of every part of the process; it's personal to us. But there is lots that we don't have to do. (Wright-Patterson AFB Focus Group 2)

Using reachback may not affect these problem categories, however. In the next chapter, we recommend some policy changes that will.

CHAPTER SIX

Conclusions and Recommendations

The AAR, focus group, and JCCS contract data provide a broad picture of what CCOs have done in recent contingencies, how their purchases are distributed using different contracting approaches, and the challenges CCOs face in accomplishing their mission because of tasks that might interfere with their "real" job of purchasing goods and services for the warfighter. While some of these challenges may be mitigated by the use of reachback, others can be addressed through changes in procedures and policies.

CCO Challenges and Approaches to Meeting Them

Analysis of JCCS data and focus group responses suggests the importance of the following:

1. **Facilitating better requirements/SOW development.** Inexperienced customers in-theater do not know how to write requirements or SOWs, so CCOs are often burdened with writing them. The latest online copy of the joint contingency contracting handbook (DPAP, 2009) has an appendix with a SOW checklist and a collection of sample SOWs, but better support could include remotely accessible help desks, similar to the one operated 24 hours a day, seven days a week by AFCESA to connect personnel with various civil engineering questions to the appropriate in-house expert (AFCAP interview, February 26, 2009; Cuttita, 2007).

2. **Consolidating requirements to fewer contracts.** We saw in Chapter Two that, in FY 2008, CCOs in JCC-I/A wrote more purchase order contracts and actions than any other type, with most of those contracts used once. By promoting greater consolidation of requirements across customers and locations, it would be possible to reduce the number of contracts. To do this, the Air Force must develop ways to make it easier for CCOs to access active contracts that could meet a requirement. Databases or other information repositories should also be developed that allow CCOs to find contracts that might be useful to them but are written by other organizations that are near them geographically. As we heard in one of our focus groups, "There was a general lack of knowledge of knowing what contracts were there in-theater that we could use. Biggest surprise was that people don't know what's out there as far as existing contracts AOR-wide" (Shaw AFB Focus Group 1).

3. **Revising deployment policies.** As described in Chapter Five, focus group comments made it clear that CCOs believe there are some basic contract tasks that can be accomplished in-theater by lower-grade personnel and other tasks that could be accomplished

by personnel without warrants. Allowing the deployment of lower-grade personnel increases the available deployment pool and also provides training for inexperienced personnel. Allowing personnel such as those in the 63A (acquisition manager) career field to deploy to positions that are currently filled by contracting personnel but that do not require a warrant would also increase the deployable pool of personnel and allow people with warrants to concentrate on the business of buying. It could also bring in expertise for "big" buys. Focus group participants also suggested that more use could be made of civilian contracting personnel in lower-threat environments, such as Kuwait and Qatar.

4. **Reviewing personnel allocations periodically and revising as needed.** We saw in Chapter Five that several CCOs experienced situations in which their offices had too few people. There have also been locations that had too many personnel for the work required—with undesirable consequences:

> The captain I replaced in Kuwait was cleaning the file room. A CCO—an O3—spent 90 days cleaning the file room. How do you justify that? Why send her there? She's separating; she said it wasn't worth it.

In a related vein, one office had a higher proportion of highly skilled personnel than it needed, whereas another had too low a proportion. Accordingly, we recommend that workload metrics (e.g., transaction volume per CCO) and CCO comments should be used periodically to revise personnel allocations. For joint positions, this will require negotiation with other services, and evidence of stress on the Air Force's contracting workforce (e.g., changes in accessions or attrition rates) would bolster the Air Force's position. In the cases of Kuwait and Qatar, whose CCO contingents could be reduced, according to focus group participants, the Air Force can make changes on its own.

5. **Clarifying the roles of other personnel in the contracting process.** This includes individuals responsible for performance monitoring (e.g., CORs and QAEs), as well as those responsible for escort duty and base access for vendors. While in a deployed setting, personnel often spend time on tasks outside their AFSC, but CCOs reported situations as extreme as arguing with customers over responsibility for a task like escort duty. A greater understanding of contracting roles and responsibilities on the part of the customer and other military personnel may lead to the designation of more-appropriate personnel to serve as CORs (i.e., people who already possess the required expertise or who have been better trained) or the allocation of time to assist in vendor base access–related tasks.

Reachback Potential

By providing continuity, standardization, training, and a repository of expertise (as shown in Table 4.5), reachback has the potential to improve the efficiency and effectiveness of contracting. Using reachback for some purchases also has the potential to reduce the number of deployed Air Force contracting personnel. Our rough analysis, based on the purchase categories that contracting personnel believe are amenable to reachback and the actions (or dollars) spent in these categories, indicates that the reductions could be significant. However, reducing the number enough to change the current 1:1 deploy-to-dwell time will be a challenge because of the number of joint billets involved. We recommend that the Air Force do the following:

1. **Refine these estimates**—for example, analyze purchase categories more closely to determine where they overlap. Also, the changing nature of purchases as a conflict moves from the buildup phase to the sustainment phase suggests that deployments are more important in the initial phases of a contingency, and reachback becomes more of an option as the transition is made to the sustainment phase.

2. **Further analyze the experiences of existing reachback organizations (AFCEE, AFCAP, 20 CONS, and the JCC Reachback Cell at RICC) to develop lessons learned about the challenges of managing contracts from a distance.** Although all these organizations will provide useful lessons, the JCC Reachback Cell, because of its recent establishment and the fact that it was specifically designed to work in a joint environment, may be a particularly useful model for how a reachback organization should be organized, how large it should be, and how it should interact with the customers (and commanders) in-theater. Learning from the experiences of other organizations is especially important because, as we have seen, potential benefits of reachback include improved strategic buying and the concentration of expertise for larger source selections.

For Further Research: Reachback Implementation Issues

There are several issues related to implementing reachback that were beyond the scope of the current effort but emerged during our study as important for further research.

1. **Joint and command-and-control issues.** More than 70 percent of current USAF deployments for contracting personnel are joint. What will the Air Force need to do to convince participants in the joint manning document process that Air Force deployments can be reduced using reachback? Focus group participants frequently mentioned the importance of a commander knowing that a CCO worked for him or her and that the CCO could be depended on to satisfy the commander's needs. What command-and-control relationship should exist between members of a reachback cell and the in-theater commander?

2. **Policy issues.** The use of reachback may affect purchases in unexpected ways. For example, the simplified acquisition threshold for an overseas contingency is $1 million.[1] For a contingency in CONUS, it is $250,000, and under normal circumstances, it is $100,000. Which threshold applies for a CONUS-based reachback cell making purchases for an overseas contingency? Using SAP can reduce administrative costs and decrease the time needed to make a purchase; if reachback is used more extensively, it would be beneficial if the reachback cell could operate under contingency thresholds.[2]

---

[1] Additionally, simplified acquisition procedures are authorized for acquisitions that do not exceed $11 million when the acquisition is for commercial items that are to be used in support of a contingency operation (see FAR 13.500(e)(1)).

[2] Without SAP, the reachback cell could not be as responsive as required by commanders in-theater. Policies and practices designed to maximize opportunities for full and open competition and for small businesses or to conform to DoD, Air Force, or other services' contracting policies would add time to the process and slow contract awards.

The Air Force and other services might need to seek a waiver for SAP threshold limits if reachback is used for more contracting.

3. **Collateral impact.** What impact will the establishment of a reachback approach to some contingency contracting have on training requirements or career development? For example, we noted that some focus group participants felt it would be good to have experience in a reachback cell before being deployed, whereas others argued that only people who had been deployed should be in a reachback cell. We have also seen that formal or strategic source selections are becoming more important in the sustainment phase of contingencies, but contracting officers who have done only support contracting are not getting the experience they need to manage such selections. Will the use of reachback mean that new skills are required?

4. **Reachback cell location and personnel pool.** The JCC Reachback Cell is located in Illinois, Shaw AFB is in South Carolina, JCC-I/A has a closeout center in San Antonio, and we heard in focus groups that contract review for OIF contracts was done in San Francisco. What are the advantages and disadvantages of these locations? Considerations include facility availability issues, communication issues (related to time zones), and manpower issues related to assigning personnel to a reachback location. Is it best to have a reachback location in the United States, or should a location closer to, but not in, the theater be considered?

5. **Resources.** CCOs stressed the importance of a communication infrastructure that allows around-the-clock telephone and email contact and file-sharing capability. Can locations in-theater maintain reliable connectivity with a reachback facility? How will ensuring this capability affect the cost of a reachback facility? What other resources (office space, equipment) will be needed for such a facility?

Our research has shown that analysis of a variety of rich data sources—AARs, focus groups, interviews, and contracting data—can provide insight into not only the potential use of reachback to decrease deployment demands for Air Force CCOs but also into issues unrelated to reachback that contribute to career field stresses. These same data sources will be valuable in addressing potential challenges related to implementation of reachback and nonreachback policies designed to relieve stress on the Air Force's contracting personnel.

General Information About AFCAP, AFCEE, and the JCC Reachback Branch

AFCAP

AFCAP is the Air Force's tool for meeting "extreme situation, massive response" requirements for small-scale contingency operations, humanitarian operations, and base operating support for military operations other than war.[1] Commanders or any other government agency can use it for vastly ramping up or down base operation support services worldwide, including within the United States. The program is managed by the Air Force Civil Engineer Support Agency (AFCESA) at Tyndall AFB, Florida, and the Air Force Services Agency at San Antonio, Texas. AFCESA is a field operating agency that reports to the Office of the Civil Engineer of the Air Force under AF/A4/7 at the Pentagon (USAF, undated).

AFCAP was established in 1997 as the Air Force drew down its active-duty force and needed to provide surge civil engineering service support in response to emergencies. Its focus is on services or temporary construction, not large-scale permanent construction, which falls under AFCEE. Some of the services AFCAP offers are carpentry, plumbing, electrical, mechanical, air conditioning, food service, lodging management, laundry plant operation, fire protection emergency management, professional engineering, and project and program management.

Originally a five-year, $475 million program, the most recently awarded AFCAP III contract is for ten years and has a maximum program ceiling of $10 billion, making it the largest contingency support contract awarded by the Air Force. It has line items that are cost-plus-award-fee, cost-plus-fixed-fee, and firm-fixed-price. Each AFCAP firm is guaranteed at least $15,000 over the life of the contract.

After receiving a customer requirement with funding, AFCESA competes the task order and writes a contract for the project. Contracting officers and AFCESA program managers are located at Tyndall AFB. Some task orders have AFCAP contract program managers who are located in the AOR.[2] The customer is expected to provide contract administration, quality assurance, and task order surveillance (COR responsibilities). AFCAP charges a fee for service to its non–Air Force customers (U.S. Government Accountability Office, 2004).

Competed in 2005 among firms capable of providing a broad array of civil engineering services, either separately or jointly with partner firms, AFCAP contracts were awarded in November 2005 to six firms qualified to compete for task orders that might arise for specific projects and requirements. These firms are Washington Group International, CH2M Hill

---

[1] USAF, 2006. See also "$10B AFCAP Program," 2005.

[2] Interview at AFCESA, Tyndall AFB, Florida, February 26, 2009.

Global Services, URS/Berger JV (Joint Venture), Bechtel National, DynCorp International, and Readiness Management Support (RMS).[3]

Previous AFCAP contracts had been sole-source contracts. The first AFCAP cost-reimbursement award-fee contract was awarded to RMS in February 1997.[4] AFCAP II was awarded in February 2002 again to RMS under a cost-plus-award-fee contract (Elliott, 2006). Prior to 1997, the kinds of capabilities provided by AFCAP were assumed by active-duty and reserve Air Force personnel.

Some of the notable uses of AFCAP have been to support U.S. forces in Kosovo, humanitarian efforts after the tsunami in the Indian Ocean in 2004, relief and recovery operations after Hurricanes Katrina and Rita in the summer of 2005, the earthquake in Pakistan in October 2005, the rebuilding of Iraq (including the restoration of water and power to Baghdad) in 2003, operations in Afghanistan, and relief efforts in the Horn of Africa. In the early part of the contingency in the AOR, AFCAP was used for air traffic management at air bases throughout central Asia (Walker, 2004). Later, it was used for power production and professional engineers, war reserve materiel equipment augmentation, heavy equipment leases, rock quarry operations and concrete batch plants, and to fortify border crossings.[5]

AFCEE

AFCEE, located at Brooks City-Base, San Antonio, Texas, provides engineering and environmental services to Air Force and joint installations. In 1991, during an early wave of manpower reductions, technical and professional services in environmental and installation planning and engineering, and military housing construction and privatization were consolidated at Brooks AFB (Coalition Provisional Authority, 2004). In 1997, AFCEE further consolidated and assumed all of the Air Force's military construction, including contingency construction (Dominguez, 2008). AFCEE is responsible for "sustainable" or permanent construction peculiar to Air Force requirements falling outside the U.S. Army Corps of Engineers, which amounts to about 4 percent of the FY 2008 DoD military construction and 74 percent of environmental restoration dollars (Firman, 2009). As a field operating agency, AFCEE's chain of command is the Office of the Air Force Civil Engineer and AF/A4 in the Pentagon.

AFCEE supports the Multi-National Coalition–Iraq and Multi-National Security Transition Command–Iraq in planning, construction oversight, and environmental cleanup (Coalition Provisional Authority, 2004). It also supports the Combined Security Transition Command–Afghanistan to reconstitute the Afghan National Air Force. For example, AFCEE is supporting the buildup of Bastion Air Base in Helmand Province, Afghanistan, by extending runways to accommodate U.S. forces joining British forces.

AFCEE contracts are written and managed in CONUS. Its contracting officers are in CONUS, although its program managers are located in the AOR.[6] The AFCEE contractor provides quality assurance and construction surveillance (DoD Inspector General, 2008). The

---

[3] Their contract numbers are FA3002-06-D-0001 to FA3002-06-D-0006, respectively.

[4] Contract number F08637-97-C-6001.

[5] *Smart Book for CONCAP, LOGCAP, AFCAP*, undated. See also Elliott, 2006.

[6] Interview with AFCEE personnel, Brooks-City-Base, Texas, February 12, 2009.

customer provides contract administration and task order surveillance (COR responsibilities). AFCEE charges a fee for services for its non–Air Force customers.

AFCEE Worldwide Environmental Restoration and Construction (WERC) contracts have been used to construct hundreds of installations in Iraq and Afghanistan (Coalition Provisional Authority, 2004; Eulberg, 2009). Current WERC contracts are the fifth in a series of IDIQ environmental and construction contracts. They have a program ceiling of $4 billion and, if necessary, can increase to $10 billion. In November and December 2003, 40 contractors bid on WERC contracts and only 27 were awarded them. Each of the 27 WERC contracts had the same terms and conditions and differed only in the contractor's rate structure and various partners and subcontractors. Four of the 11 large WERC contractors were awarded sector contracts by the CPA. Sixteen of the 27 contractors are small businesses.[7]

AFCEE competes requirements among these 27 contractors and awards task orders to winning contractors. The FAR permits other government agencies to order from these contracts.

The CPA used WERC contracts to meet high-priority goals of completing numerous structures for the New Iraqi Army by June 2004. The AFCEE WERC contract fit the need: It was an active contract, it had originally been competed and awarded to contractors deemed competent to perform a variety of construction and environmental restoration activities, and it included small businesses.

JCC Reachback Branch

The JCC Reachback Branch was established by a memorandum of agreement with JCC-I/A on January 5, 2009. It is in the Reachback Division of the Rock Island Contracting Center in Illinois, which is under the Army Contracting Command (ACC).[8]

The JCC Reachback Branch also supports the Surface Deployment and Distribution Command (SDDC), the Army's service component to the U.S. Transportation Command. SDDC has responsibility for all land transportation and port operations. Another branch in the Reachback Division, the Southwest Asia Support Branch, focuses on Kuwait and Qatar.

Employees in the RICC are Department of the Army civilian contracting officers. One military officer is located forward at JCC-I/A to serve as a liaison with the JCC Reachback Branch; a civilian liaison officer for the branch is located in Afghanistan. The purpose of the reachback branch is to "reduce the JCC-I/A workload, provide personnel continuity . . . and assist in identifying and executing strategic sourcing candidates" (RICC, 2009).

The decision to use the JCC Reachback Branch for contract actions is made jointly by the commander of JCC-I/A and the RICC director; once the decision is made, the Reachback Branch provides the "full range of acquisition support, from acquisition planning through contract closeout" (RICC, 2009). This support includes, but is not limited to, acquisition planning; all necessary pre-award procurement policy and legal reviews; pricing; solicitation

---

[7] WERC contracts have provisions for firm-fixed-price, fixed-price-incentive, cost-plus-fixed-fee, cost-plus-incentive-fee, and cost-plus-award-fee pricing.

[8] Before October 2008, RICC was under the Army Sustainment Command (JCC-I/A branch interview, March 20, 2009). ACC was formally established on October 1, 2008. ACC is a major subordinate command of AMC. ACC performs the majority of contracting work for the U.S. Army. It consists of two subordinate commands responsible for installation and expeditionary contracting and other Army contracting elements. See ACC, undated. The executive director of LOGCAP is also located at RICC.

preparation and issuance; proposal evaluation; negotiations; and contract preparation, award, post-award contract administration, and contract closeout (RICC, 2009).

Analysis Methodology for CCO After Action Reports

As part of our study, we analyzed CCOs' firsthand accounts of their deployment experiences, as documented in AARs. As per AFFARS Appendix CC, within 30 days of returning from a deployment, CCOs are required to submit to their parent MAJCOM an AAR that covers the following topics:[1]

- site survey update, as needed
- problems encountered at the site with the contracting process
- availability of local transportation and billeting
- availability of communication resources
- evaluation of any host-nation agreements or status of forces agreements as they pertain to contingency contracting
- adequacy of facilities and resources provided
- specific problems that could be anticipated to support an extended exercise or contingency operation at the location
- other requirements related to rank, gender, skill set, contingency kit, or individual supplies.

The Contingency Contracting section of the website maintained by the Defense Acquisition University serves as an access-restricted repository for unclassified AARs so they can be reviewed by CCOs, acquisition community leadership, and other proponents of the acquisition community. In conducting this study, we requested and received access to the reports, and in January 2009, we downloaded all AARs available on the website.[2] Once duplicate reports were eliminated, we arrived at a total of 236 AARs to include in our dataset. The AARs spanned countries and contingencies (e.g., exercises, humanitarian aid, war).

We reviewed a random subset of the AARs (approximately every fifth AAR in our dataset) to develop a list of topics that were relevant to the goals of the study. This was both a deductive and inductive process; while some topics were obvious given study parameters, the extent to which they may have been addressed in the AARs—if at all—was not clear at the outset of our analysis. In addition, this preliminary review of a subset of the AARs suggested other topics of potential value that were not initially apparent.

We organized these topics into a coding "tree" to facilitate tagging relevant interview excerpts. A coding tree is a set of codes, the "labels for assigning units of meaning to informa-

---

[1] USAF, 2007, para. CC-502-4, "Contingency contracting activity during termination/redeployment."

[2] Although we included all of the AARs available on the Defense Acquisition University's website, it is unclear whether they constitute all of the AARs prepared by CCOs during the time frame of our analysis or only a subset. Accordingly, the results of our analysis should not be considered representative of the entire set of CCO deployment experiences.

tion compiled during a study" (Miles and Huberman, 1994, p. 56). Codes are used in the data reduction process to retrieve and organize qualitative data by topic and other characteristics. The coding tree used in the analysis of the AARs included the following topics:

- AAR author attributes (e.g., CCO service, rank)
- deployment characteristics (e.g., location, year)
- task descriptions
- workload issues
- contracting problems
- transition issues
- successes and best practices
- recommendations.

To facilitate qualitative analysis of the AARs, we transformed them from various software formats into simple text files (.txt) and removed exhibits such as photographs, maps, and vendor lists. Our focus was on the narrative portion of each AAR—the CCO's actual account of his or her deployment experience. We opted to use QSR N6 for our analysis of the AARs. N6 is a software package that enables its users to review, categorize, and analyze qualitative data, such as text, visual images, and audio recordings. Software like N6 permits analysts to assign codes to passages of text and then later retrieve passages of similarly coded text within and across documents. N6 is also capable of simple word-based searches as well as more sophisticated text searches, such as Boolean searches involving combinations of codes.

Coding AAR author attributes and characteristics of the deployment, such as location, was a relatively straightforward process because these are factual, objective data. Ninety-four percent of the AARs (221) were authored by Air Force personnel, and the majority of them (51 percent, 121 AARs) were written by enlisted personnel. The remaining AARs were authored by officers (27 percent, 64 AARs), civilians (1 percent, 3 AARs), or a combination of personnel (20 percent, 48 AARs). In addition, a breakdown of the 236 AARs by location and the year the deployment began is provided in Table B.1. The largest numbers of AARs were accounts of deployments to Iraq (38 percent, 89 AARs) and Afghanistan (14 percent, 33 AARs). Timing-wise, 27 percent of the AARs (63) discussed deployments that began in 2004, and 20 percent of them (48) pertained to deployments that commenced in 2006. Taking a closer look by country, 55 percent of the AARs based in Iraq covered deployments beginning in 2004–2005, and 39 percent of the AARs based in Afghanistan described deployments that began slightly later, in 2006. All in all, these descriptive statistics indicate that the dataset featured a relatively high level of variation in terms of both author traits and deployment features.

Reviewing the AARs and coding them for more substantive or subjective codes, such as task descriptions and contracting problems, was more time-intensive and involved all three authors of this report. An iterative process of coding a set of AARs, sharing examples of coding for validation, and making refinements as needed was used to ensure that the original set of codes was applied to the text in a manner with which all three researchers agreed. Coding passages related to task descriptions is a good example of this approach: Not only did we initially collaborate to arrive at a series of steps or phases by which to categorize the contracting process, but we also had discussions informed by a review of AAR passages to determine where, within the

Table B.1
AAR Descriptives: Deployment Characteristics

Country	Year Deployment Began											Total by Country
	N/A	99	00	01	02	03	04	05	06	07	08	
Afghanistan	1	0	0	0	0	4	5	2	13	7	1	33
Australia	0	0	0	0	0	2	1	0	0	0	0	3
Bosnia	0	0	0	0	0	1	1	1	0	0	0	3
Cambodia	0	0	0	0	1	1	0	0	0	0	0	2
Diego Garcia	0	0	0	0	0	0	0	3	1	0	0	4
Guam	0	0	0	0	0	1	0	0	0	0	0	1
Horn of Africa	0	0	0	0	0	0	0	0	2	0	0	2
Honduras	0	0	1	0	0	0	0	0	1	0	0	2
India	0	0	0	0	0	0	1	0	0	0	0	1
Indonesia	0	0	0	0	0	0	1	0	0	0	0	1
Iraq	1	0	0	0	0	8	29	20	18	10	3	89
Jordan	0	0	0	0	0	0	2	0	0	0	0	2
Korea	1	0	0	2	1	1	1	0	0	0	0	6
Kosovo	0	0	0	0	0	0	2	1	1	0	0	4
Kuwait	0	0	0	0	1	1	4	0	2	0	0	8
Kyrgyzstan	0	0	0	0	0	1	1	0	1	1	0	4
Laos	0	1	0	0	0	0	0	0	0	0	0	1
Malaysia	0	0	0	0	0	0	1	0	0	0	0	1
Mongolia	0	0	0	0	0	1	0	0	0	0	0	1
Morocco	0	0	0	0	0	0	0	0	0	1	0	1
Oman	0	0	0	0	2	0	0	0	0	0	0	2
Pakistan	0	0	0	0	0	1	1	0	1	0	0	3
Philippines	1	0	1	0	5	4	2	1	1	0	0	15
Qatar	0	0	0	0	1	2	1	0	3	1	0	8
Romania	0	0	0	0	0	1	0	2	0	0	0	3
Saudi Arabia	0	0	0	0	0	0	0	1	1	0	0	2
Southeast Asia	0	0	0	0	0	1	1	2	0	0	0	4
Thailand	0	0	2	2	2	2	3	0	1	2	0	14
Turkmenistan	0	0	0	0	0	0	2	0	0	0	0	2
United Arab Emirates	0	0	0	0	0	1	2	1	2	1	1	8
Uzbekistan	0	0	0	0	0	0	2	2	0	0	0	4
Multiple countries	0	0	0	0	0	1	0	1	0	0	0	2
Total by year	**4**	**1**	**4**	**4**	**13**	**34**	**63**	**37**	**48**	**23**	**5**	

SOURCE: CCO AARs (N = 236).

contracting process, certain types of tasks should be allocated. Similarly, while perusing the AARs to apply the initial set of agreed-upon codes, when patterns emerged that suggested the need for new codes, we worked to achieve consensus on what the new codes should include and how they may inform the analysis.

After all the AARs were coded, coding reports were run to reveal the frequency with which each code was applied. Additional reports were generated so that all the passages coded with a specific code could be reviewed together. Finally, since the same passage could be assigned many different codes (as applicable), Boolean searches were performed so that both frequencies and specific coding examples could be reviewed for intersections of codes (i.e., passages coded with two different codes, such as a specific task description and as a problem type). Analyses of these reports were then considered in conjunction with other data sources (e.g., interviews, JCCS contracting data) to answer research questions key to the project's goals.

Joint Contingency Contracting System Data

Joint Contingency Contracting System

Contract data came from two sources: the JCCS, which was the source for most of the contract data analyses, and the FPDS-NG.[1] JCCS, as a contract action data system concept, was established in November 2006 when the JCC-I/A commander's contracting authority was extended to all contracting activities assigned or attached to USCENTCOM (with the exception of the U.S. Army Corps of Engineers [U.S. Central Command, 2006]). JCCS operations began in FY 2007. The system is used to record contract information and to facilitate contracting with host-nation vendors by enabling them to register online (similar to the Central Contractor Registry in the United States), review posted solicitations, and send bids (see JCCS, 2009).[2] It has been deployed to 20 Iraq and six Afghanistan regional contracting centers (RCCs), reconstruction offices at JCC-I/A headquarters in Baghdad, Special Operations Command headquarters in Tampa, Florida, and nine international offices. Earlier contracting actions in Afghanistan and Iraq were recorded and stored on computers by CCOs at their respective RCCs. Historical data prior to October 2006 were collected from these locations and stored in JCCS, though their completeness and accuracy cannot be verified.[3]

We downloaded JCCS data several times during the study. Our analyses are based on data downloaded on January 22, 2009. The data variables analyzed in the JCCS were

- contract_type
- RCC
- division
- PIIN (procurement item identification number)
- dollars
- awd_date
- is_host_nation_bus
- contract class.

A variable created called "country" was inferred from the RCC and division. It took on a value of either "Iraq" or "Afghanistan." The PIIN is the contract number, which normally

---

[1] For more on FPDS-NG, see FPDS-NG, undated. These data were downloaded April 18, 2009.

[2] JCCS received a Computerworld Honors Program award in 2008. See Computerworld Honors Program, 2008.

[3] Email received from JCC-I/A on February 11, 2009.

has 13 alphanumeric characters.[4] Fiscal year was verified by the award date, "awd_date." The variable "is_host_nation_bus," which indicates whether the provider is a local business, was either "0" or "1." "Contract class" had values of "commodity," "construction," "services," or "unknown," if the information was missing.

Contract type was determined by using the "contract_type" variable and/or the "PIIN" variable. If "contract_type" was "GPC" or "government purchase card," then the contract type was assumed to be "GPC," since GPC cards can be used as either a contract or to make a payment on an existing contract. If contract type was "SF44," then the contract type was assumed to be as given. If the PIIN took the traditional form (13 positions, etc.) and contract_type was not "GPC," then contract type was inferred directly from the ninth position of the PIIN. If the PIIN was missing, the value in the PIIN did not fit traditional contract nomenclature, or the PIIN was a federal supply or government services schedule contract, then contract type was also inferred by the value in the variable "contract_type."

FPDS-NG data for contracts used by DoD organizations were also analyzed for work performed in the AOR in FY 2008 to determine how much was being spent on external support contracts written outside Iraq and Afghanistan. Federal contract action data of $2,500 or more are recorded in FPDS-NG.[5] Contracts such as AFCAP and AFCEE appear in FPDS-NG and not in JCCS. The FPDS-NG data analyzed were downloaded on April 18, 2009.

The variables used in the FY 2008 FPDS-NG data analyses were

- PIID or IDVPIID (procurement instrument identifier and indefinite delivery vehicle procurement instrument identifier)
- dollars
- number of actions
- place of performance, country
- effective date (the date the contract became active).

All DoD data were analyzed for purchases made where the work was performed in Iraq, Afghanistan, Kuwait, United Arab Emirates, and Qatar. Purchases made for work performed elsewhere in USCENTCOM included countries in which several different kinds of missions were being conducted, such as Turkey and Saudi Arabia. For this reason, they were not included.

Detailed Data Tables for Figures in Chapter Two

Tables C.1 and C.2 contain the data on which Figures 2.1 through 2.3 are based.

---

[4]   Positions 1 through 6 of the PIIN identify the department or agency office that issues the contract. Positions 7 and 8 are the last two digits of the fiscal year, and position 9 indicates the type of contract.

[5]   Beginning in FY 2005, FAR 4.602(c)(1) requires contract actions of at least $2,500 to be recorded in FPDS-NG. Before that, the threshold was $25,000.

Table C.1
FY 2008 Work Performed in USCENTCOM by Country: Total DoD Dollars, Contracts, and Actions

Country	Dollars (millions)					Percentage	
	JCC-I/A and FPDS	JCC-I/A Only	FPDS Only	Total JCC-I/A	Total	Total JCC-I/A	Total JCC-I/A, FPDS
Iraq	5,221	759	10,005	5,980	15,985	80	57
Afghanistan	1,232	306	4,898	1,537	6,435	20	23
Kuwait	n/a	n/a	4,121	n/a	4,126	n/a	15
UAE	n/a	n/a	1,082	n/a	1,082	n/a	4
Qatar	n/a	n/a	328	n/a	328	n/a	1
Total	6,452	1,065	20,434	7,517	27,957	100	100

Country	Contracts					Percentage	
	JCC-I/A and FPDS	JCC-I/A Only	FPDS Only	Total JCC-I/A[a]	Total JCC-I/A, FPDS[a]	Total JCC-I/A[a]	Total JCC-I/A, FPDS[a]
Iraq	14,228	2,978	3,908	17,206	21,114	64	62
Afghanistan	8,352	1,327	1,407	9,679	11,086	36	33
Kuwait	n/a	n/a	1,040	n/a	1,040	n/a	3
UAE	n/a	n/a	458	n/a	458	n/a	1
Qatar	n/a	n/a	189	n/a	189	n/a	1
Total[a]	22,575	4,305	6,924	26,880	33,804	100	100

Country	Contract Actions					Percentage	
	JCC-I/A and FPDS	JCC-I/A Only	FPDS Only	Total JCC-I/A	Total JCC-I/A, FPDS	Total JCC-I/A	Total JCC-I/A, FPDS[b]
Iraq	21,574	4,504	6,450	26,078	32,528	63	49
Afghanistan	13,407	1,948	2,783	15,355	18,138	37	27
Kuwait	n/a	n/a	12,850	n/a	12,850	n/a	19
UAE	n/a	n/a	2,055	n/a	2,055	n/a	3
Qatar	n/a	n/a	799	n/a	799	n/a	1
Total	34,981	6,452	24,937	41,433	66,370	100	100

SOURCES: FY 2008 JCCS data, downloaded from the JCC-I/A website on January 22, 2009; FY 2008 FPDS-NG data were downloaded on April 18, 2009.

NOTES: Common contracts are in the category "JCC-I/A and FPDS"; contracts only in the JCC-I/A data are in the category "JCC-I/A Only"; and contracts only in FPDS are in the category "FPDS Only." The column "Total JCC-I/A" is the sum of the categories "JCC-I/A and FPDS" and "JCC-I/A Only" and associated respective percentages. The column "Total" is the sum of the first three columns.

[a] Contract totals can be less than the sum of the elements in each category because some contracts, such as those for theater-wide requirements, are used in more than one country.

[b] Percentages might not sum to 100 due to rounding.

Table C.2
FY 2008 JCC-I/A Dollars Spent, Number of Contracts and Actions, by Commodity Class and Vendor Type

| Commodity Class | Dollars (millions) | | | Percentage |
	Nonlocal Vendors	Local Vendors	Total	Total
Service	3,105	1,069	4,174	56
Commodity	707	1,150	1,857	25
Construction	346	1,137	1,483	20
Unknown	1	3	3	0.04
Total	4,159	3,359	7,517	100

| Commodity Class | Contracts | | | Percentage |
	Nonlocal Vendors	Local Vendors	Total[a]	Total[a]
Service	1,081	3,817	4,846	18
Commodity	9,940	9,527	19,403	72
Construction	356	2,540	2,884	11
Unknown	5	5	10	0.04
Total[a]	11,314	15,706	26,880	101

| Commodity Class | Actions | | | Percentage |
	Nonlocal Vendors	Local Vendors	Total	Total
Service	3,292	7,106	10,398	25
Commodity	12,028	14,216	26,244	63
Construction	706	4,068	4,774	12
Unknown	6	11	17	0.04
Total	16,032	25,401	41,433	100

SOURCE: FY 2008 JCCS data, downloaded from the JCC-I/A website on January 22, 2009.

NOTE: Nonlocal vendors are those not headquartered in Iraq or Afghanistan. Local vendors are headquartered in either Iraq or Afghanistan. Missing commodity class information is noted as "unknown."

[a] Contract totals are less than the sum of each commodity class because some contracts are coded as having nonlocal and local vendors and more than one type of commodity class. Total percentages exceed 100 because some contracts are coded with more than one type of commodity class or vendor.

Protocol Used for CCO Focus Groups

Focus Group Selection

Focus groups were conducted with CONUS-based military personnel who had recently returned from a deployment (typically within the past year) as well as with civilian personnel who supported contingency operations in an informal reachback capacity. To ensure that sufficient personnel were available to participate in focus groups, we worked with our research sponsors to identify Air Force bases in CONUS with a relatively large number of contracting officers. This proved a somewhat difficult task, in part because a unit is typically supported by only a small number of contracting personnel and in part because so many personnel were deployed in-theater during the time frame available for data collection. Ultimately, the locations we identified as most suitable for focus groups included Hill AFB, Utah; Wright-Patterson AFB, Ohio; Lackland and Randolph AFBs in Texas; and Shaw AFB, South Carolina.

At each location, a local POC was identified to assist the project team in scheduling focus group sessions. The POC worked with local commanders to identify a date that would work for as many contracting personnel as possible given training, planned leave, and other events (such as changes in command), and he or she worked with the RAND team to populate the focus groups as well. All local, eligible contracting personnel received a request to participate that included a description of the study's objectives and more-specific details about the focus group sessions, including topics covered, location, date, and time. To the extent possible, focus groups were stratified so that only one type of personnel—officers, enlisted personnel, or civilians—constituted a group. We learned during the focus groups that most individuals who served as CCOs in a contingency setting did so in either Afghanistan or Iraq. Many of them had multiple experiences as CCOs, spanning different parts of Afghanistan and Iraq as well as other parts of southwest Asia (e.g., Qatar) and points beyond USCENTCOM (e.g., Bosnia). Although the focus groups were not based on a random sample and their results are not generalizable to the entire contracting workforce, a broad range of experiences was represented in the sessions. Contracting personnel not only served in a variety of locations but also supported different types of customers (e.g., the Army, JCC-I/A) and held different positions in-theater.

Protocol

1. We're going to begin by going around the room. Please each take a turn and tell us briefly about four things: your current paygrade; how long you've been in the Air Force and, if different, how long you've been in the contracting career field; and your deploy-

ment experience, specifically where and when you've supported any type of contingency since 9/11.

2. We're very interested in learning about the contracting-related tasks you performed on your most recent deployment. What were your primary responsibilities, and what tasks took up most of your time?

Probes

- Which tasks were most likely to get done, regardless of how busy you were?
- On the flip side, which tasks tended to fall along the wayside because of how busy you were?
- Which did not necessarily need a CCO to handle them? Which *could or should have* been performed by someone else?

3. What types of contracting-related problems did you encounter while deployed, if any?

4. During your deployment, did you have any experience with external support contracts, such as AFCAP or LOGCAP? How so?

5. I asked about external support contracts like AFCAP because they are current examples of reachback. More generally, how have you seen reachback used to support contingency contracting? We're interested in both formal and informal applications of reachback.

6. Which contracting tasks do you think would be best suited to reachback? In other words, which tasks could have been performed elsewhere, or would even have been better performed elsewhere? Why?

   a. Which absolutely needed to be handled in-theater, and why?

Probes (as needed)

- What about the development of requirement packages or statements of work? Is either task amenable to reachback? Why or why not?
- Would source selection be amenable to reachback? Why or why not?

7. Are there certain types of *requirements* that are well-suited to reachback? Which ones, and why?

   a. Are there types that most likely are not—that are better handled in-theater? Why?

   b. Does it matter whether the requirement is related to commodities, services, or construction? [If needed: For example, are certain types of commodities better purchased outside the AOR, or must all tasks for services be performed in-theater?]

   c. We've heard mixed views about whether complex requirements are good candidates for reachback or should be handled in-theater. In your view, what makes a requirement complex? How might that affect its suitability for reachback?

8. We've discussed types of tasks and requirements as potential criteria or parameters that could be used to decide whether to rely on in-theater contracting support or to use some form of reachback. What other criteria or parameters could guide this decision?

Probes (as needed)

- How do local socoeconomic requirements affect the potential for reachback, if at all?
- Would reachback be more feasible when the threat level is low, or are there opportunities to use reachback even in a hot environment?

9. One of the perceived benefits of reachback is it would reduce some of the stress on CCOs. Do you think this is true? Why or why not?

10. What [other] benefits might be obtained with greater use of reachback?

11. What are potential downsides of reachback or reasons why some individuals—be they customers, JCC-I/A leadership, or other personnel—would oppose reachback?

12. What might be ways to address those concerns?

Probe

- More broadly, how should reachback be implemented to ensure it's as successful as possible?

13. In closing, what message would you like to convey to Air Force leadership and policy-makers regarding your deployment experience in general or the prospect of reachback in particular?

References

"$10B AFCAP Program Contract Provides 'Expeditionary Engineering,'" *Defense Industry Daily,* November 10, 2005. As of January 4, 2010:
http://www.defenseindustrydaily.com/10b-afcap-iii-program-contract-provides-expeditionary-engineering-01468/

AFCEE—*See* Air Force Center for Engineering and the Environment.

AFCESA—*See* Air Force Civil Engineer Support Agency.

Air Force Association, *Air Force Almanac,* 2008. As of February 24, 2010:
http://www.airforce-magazine.com/MagazineArchive/Magazine%20Documents/2008/May%20
2008/0508facts_figs.pdf

Air Force Center for Engineering and the Environment, home page, undated. As of January 4, 2010:
http://www.afcee.af.mil/

Air Force Civil Engineer Support Agency, "Air Force Contract Augmentation Program (AFCAP) Fact Sheet," 2010. As of January 4, 2010:
http://www.afcesa.af.mil/library/factsheets/factsheet.asp?id=9381

Air Force Personnel Center, Interactive Demographic Analysis System (IDEAS), Randolph Air Force Base, Tex. As of January 4, 2010:
http://w11.afpc.randolph.af.mil/vbin/broker8.exe?_program=ideas.IDEAS_Default.sas&_service=
prod2pool3&_debug=0

Army Contracting Command, home page, undated. As of January 4, 2010:
http://www.amc.army.mil/acc/about/

Assad, Shay D., "The Department of Defense's Progress in Implementing the Recommendations of the Report of the Commission on Army Acquisition and Program Management in Expeditionary Operations," testimony before the Subcommittee on Oversight and Investigations, Committee on Armed Services, U.S. House of Representatives, March 25, 2009.

Cameron, Maj Rod A., SAF/AQCC, "Contingency: Rumors and Reality," briefing, October 27, 2008.

Chairman of the Joint Chiefs of Staff Instruction 1301.01C, "Individual Augmentation Procedures," January 1, 2004. As of June 17, 2010:
http://www.dtic.mil/cjcs_directives/cdata/unlimit/1301_01.pdf

CJCSI—*See* Chairman of the Joint Chiefs of Staff Instruction.

Coalition Provisional Authority, Office of the Inspector General for Auditing, "Audit Report: Task Orders Awarded by the Air Force Center for Environmental Excellence in Support of the Coalition Provisional Authority," Report No. 04-004, Arlington, Va., July 28, 2004. As of January 4, 2010:
http://www.sigir.mil/files/audits/cpaig_audit_afcee.pdf

Computerworld Honors Program, "DoD Business Transformation Agency (BTA), Joint Contingency Contracting System," 2008. As of June 17, 2010:
http://www.cwhonors.org/viewCaseStudy2008.asp?NominationID=2397

Correll, Roger S., USAF Deputy Assistant Secretary (Contracting), Assistant Secretary (Acquisition), "Deployment Posturing," Air Force Contracting OnPoint Memo, September 23, 2008.

Cuttita, Capt Chrissy, "Moving Forward by Reaching Back," *AFCE Magazine*, Vol. 15, No. 3, December 4, 2007. As of January 4, 2010:
http://www.afcesa.af.mil/news/story.asp?id=123077801

DCMA—*See* Defense Contract Management Agency.

Defense Acquisition University, "Contingency Contracting," undated (a). As of June 16, 2010:
https://acc.dau.mil/contingency

————, "Source Selection," undated (b). As of June 16, 2010:
https://acc.dau.mil/CommunityBrowser.aspx?id=29030

Defense Contract Management Agency, *DCMA Guidebook, Calendar Year 2009.* As of November 16, 2009:
http://guidebook.dcma.mil/18/ContRecRevconttypes.htm

Defense Federal Acquisition Regulation Supplement, Part 201, "Federal Acquisition Regulations System," July, 29, 2009. As of June 17, 2010:
http://farsite.hill.af.mil/reghtml/regs/far2afmcfars/fardfars/dfars/dfars201.htm

————, Part 213, "Simplified Acquisition Procedures," January 15, 2009. As of June 17, 2010:
http://farsite.hill.af.mil/reghtml/regs/far2afmcfars/fardfars/dfars/dfars213.htm

Defense Procurement and Acquisition Policy, *Contingency Contracting: A Joint Handbook for the 21st Century.* As of November 18, 2009:
http://www.acq.osd.mil/dpap/pacc/cc/jcchb/

DFARS—*See* Defense Federal Acquisition Regulation Supplement.

Dominguez, Gil, "The New AFCEE," *The Military Engineer*, Vol. 100, No. 651, January–February 2008, pp. 51–52. As of January 4, 2010:
http://themilitaryengineer.com/issues/January-February_2008/tme_0108.html

DPAP—*See* Defense Procurement and Acquisition Policy.

Elliott, Col Gus G., Jr., "Air Force Civil Engineer Support Agency," briefing, AFCESA, Tyndall AFB, 2006. As of January 4, 2010:
http://www.same.org/files/Members/2006ESCAFCESA.pdf

England, Gordon, Deputy Secretary of Defense, "Coordination of Contracting Activities in the USCENTCOM AOR," memorandum to secretaries of the military departments, Chairman of the Joint Chiefs of Staff, combatant commanders, and other agencies, October 31, 2006.

Eulberg, Maj Gen Delwyn R., "Building Partnerships Through Reconstruction," *Air Force Civil Engineer*, Vol. 17, No. 1, 2009.

FAR—*See* Federal Acquisition Regulations.

Federal Acquisition Regulations, Part 2, "Definitions of Words and Terms," December 10, 2009. As of February 25, 2010:
http://farsite.hill.af.mil/reghtml/regs/far2afmcfars/fardfars/far/02.htm

————, Part 13, "Simplified Acquisition Procedures," December 10, 2009. As of February 25, 2010:
https://www.acquisition.gov/far/html/FARTOCP13.html

————, Part 46, "Quality Assurance," December 10, 2009. As of February 25, 2010:
https://www.acquisition.gov/far/html/FARTOCP46.html

Federal Procurement Data System–Next Generation, home page, undated. As of June 17, 2010:
https://www.fpds.gov/

Firman, Dennis, "AFCEE Federal Opportunities Presentation," briefing, San Antonio, Tex., April 27, 2009. As of October 2009:
http://www.acec.org/advocacy/committees/pdf/anncon09_firman.pdf

FPDS-NG—*See* Federal Procurement Data System–Next Generation.

Fryer, Richard A., Jr., Commander, AFCESA, "Air Force Support for the Project Manager in the AOR," briefing, undated.

Gansler, Jacques S., Under Secretary of Defense for Acquisition, Technology and Logistics, *Urgent Reform Required: Army Expeditionary Contracting, Report of the Commission on Army Acquisition and Program Management in Expeditionary Operations* ("Gansler Commission Report"), October 31, 2007. As of July 5, 2010:
https://acc.dau.mil/CommunityBrowser.aspx?id=180366

Hageman, Maj James A., Commander, 20th Contracting Squadron, "20th Contracting Squadron Mission Brief for ACC/A7," briefing, January 16, 2009.

Holmes, Erik, "Contracting Airmen Could Deploy Every 6 Months," *Air Force Times*, October 16, 2008. As of October 6, 2009:
http://www.airforcetimes.com/news/2008/10/airforce_contracting_deployment_101408/

Hutchison, Mike, Deputy Director, Rock Island Contracting Center, "Rock Island Contracting Center Reach-Back Support to Joint Contracting Command," briefing, January 12, 2009.

JCCS—*See* Joint Contracting and Contingency Services.

Johnson, Clay III, Deputy Director for Management, Office of Management and Budget, "Implementing Strategic Sourcing," memorandum, May 20, 2005. As of June 17, 2010:
http://www.whitehouse.gov/omb/procurement/comp_src/implementing_strategic_sourcing.pdf

Joint Contracting and Contingency Services, Client Application, May 2009. As of June 17, 2010:
https://www.jccs.gov/olvr/bta_jccs_default.aspx

LeDoux, Colonel Karen E., "LOGCAP 101: An Operational Planner's Guide," *Army Logistician: Professional Bulletin of United States Army Logistics,* May–June 2005, pp. 24–29.

Long, Maj Bill, and Capt Dennis Clements, *Contingency Contracting: A Joint Handbook,* Air Force Logistics Management Agency, December 2007. [Note: See DPAP reference for an updated version of this handbook, which is now issued electronically.]

Machis, Col Denean, SAF/AQCX, "Career Field Manager's Health Assessment," draft briefing, received April 13, 2009.

Maxfield, Betty D., Chief, Office of Army Demographics, Deputy Chief of Staff of Personnel, G-1, "Army Demographics: FY06 Profile," 2006. As of February 24, 2010:
http://www.armyg1.army.mil/HR/docs/demographics/FY06%20Tri-Fold%20without%20the%20 Education%20Chart.pdf

Miles, Matthew B., and Michael Huberman, *Qualitative Data Analysis: An Expanded Sourcebook,* 2nd ed., Thousand Oaks, Calif.: Sage Publications, 1994.

Office of the Assistant Secretary of the Air Force for Contracting, *Career Planning Guide for Contracting Professionals,* July 2005. As of June 16, 2010:
http://ww3.safaq.hq.af.mil/shared/media/document/AFD-070706-066.pdf

———, *Installation Acquisition Transformation (IAT) Orientation and Status Brief,* briefing, July 2008.

Parsons, Jeffrey P., Executive Director, U.S. Army Contracting Command, record version of a statement on Army contracting before the Panel on Defense Acquisition Reform, Committee on Armed Services, U.S. House of Representatives, July 16, 2009.

Pianese, Joseph P., SAF/AQCX, "6C0X1 Historic Retention," spreadsheet, received March 23, 2009a.

———, "Department of the Air Force Civilian Retention," spreadsheet, received March 23, 2009b.

———, "Officer and Enlisted Contracting," spreadsheet, received February 25, 2009c.

Ray, Maj Butch, SAF/AQCX, "Deployment Overview: Numbers," briefing, October 2007.

RICC—*See* Rock Island Contracting Center.

Rock Island Contracting Center, *Memorandum of Agreement Between the US Army Contracting Command Rock Island Contracting Center and the Joint Contracting Command–Iraq/Afghanistan,* January 5, 2009.

Ross, Karen, "AF Contracting Competency Results," briefing, April 16, 2009.

Rump, Lt Col Nathan, SAF/AQCC, "AF Contracting Deployments as of 19 Nov 08," briefing, November 19, 2008.

SAF/AQC—*See* Office of the Assistant Secretary of the Air Force for Contracting.

SIGIR—*See* Special Inspector General for Iraq Reconstruction.

Smart Book for CONCAP, LOGCAP, AFCAP, undated. As of October 12, 2009: http://www.docstoc.com/docs/4836832/SMART-BOOK-FOR

Special Inspector General for Iraq Reconstruction, *Iraq Reconstruction: Lessons in Contracting and Procurement,* Report No. 2, July 2006.

Thompson, Gen N. Ross III, Army Director, Acquisition Career Management, "On the Army Acquisition Workforce," testimony before the Subcommittee on Oversight and Investigations, Committee on Armed Services, U.S. House of Representatives, April 28, 2009.

Thompson, Lee, Executive Director, Logistics Augmentation Program, record version of a statement before the Commission on Wartime Contracting, May 4, 2009.

U.S. Air Force, "The Air Force Civil Engineer Support Agency," fact sheet, undated. As of October 8, 2009 (password protected): https://www.my.af.mil/gcss-af/USAF/AFP40/d/1074111407/Files/editorial/AFCESAfactsheet.pdf

———, *Operational Contracting Manpower Standard,* AFMS 12A0, December 7, 2001.

———, *Personnel: Enlisted Classification,* Air Force Manual 36-2108, October 31, 2004a.

———, *Personnel: Officer Classification,* Air Force Manual 36-2105, October 31, 2004b.

———, "The Air Force Contract Augmentation Program," fact sheet, 2006. As of October 8, 2009 (password protected): https://www.my.af.mil/gcss-af/USAF/AFP40/d/1074111407/Files/editorial/AFCAP_fact_sheet_2006.pdf

———, *Air Force Federal Acquisition Regulation Supplement (AFFARS),* Appendix CC, "Contingency Operational Contracting Support Program (COCSP)," revised March 2007.

USAF—*See* U.S. Air Force.

U.S. Central Command, Fragmentary Order 09-1117, "Contracting and Organizational Changes," November 2006.

U.S. Department of Defense Inspector General, *Afghanistan Security Forces Fund Phase III—Air Force Real Property Accountability,* Report No. D-2009-031, December 29, 2008. As of January 4, 2010: http://www.dodig.mil/Audit/reports/FY09/09-031.pdf

U.S. Government Accountability Office, *Military Operations: DoD's Extensive Use of Logistics Support Contracts Requires Strengthened Oversight,* GAO-04-854, July 2004. As of January 4, 2010: http://www.gao.gov/new.items/d04854.pdf

U.S. Government Personnel Management Office, Position Classification Standard for Contracting Series, GS-1102, TS-71, December 1983. As of February 24, 2010: http://www.opm.gov/FEDCLASS/gs1102.pdf

Walker, David M., Comptroller General of the United States, *Contracting for Iraq Reconstruction and for Global Logistics Support, General Accounting Office,* testimony before the Committee on Government Reform, U.S. House of Representatives, GAO-04-869T, June 15, 2004.

Westermeyer, Colonel Roger H., PARC-I, JCC-I/A, "Joint Manning Document Brief to MNF-I CoS BG William Phillips," briefing, March 8, 2009a.

———, "JCC Reachback Weekly Update," briefing, April 27, 2009b.